Peter Robbins

Organising for Innovation

D1740364

Peter Robbins

Organising for Innovation

How leading companies accelerate innovation

LAP LAMBERT Academic Publishing

Impressum/Imprint (nur für Deutschland/only for Germany)
Bibliografische Information der Deutschen Nationalbibliothek: Die Deutsche Nationalbibliothek verzeichnet diese Publikation in der Deutschen Nationalbibliografie; detaillierte bibliografische Daten sind im Internet über http://dnb.d-nb.de abrufbar.

Alle in diesem Buch genannten Marken und Produktnamen unterliegen warenzeichen-, marken- oder patentrechtlichem Schutz bzw. sind Warenzeichen oder eingetragene Warenzeichen der jeweiligen Inhaber. Die Wiedergabe von Marken, Produktnamen, Gebrauchsnamen, Handelsnamen, Warenbezeichnungen u.s.w. in diesem Werk berechtigt auch ohne besondere Kennzeichnung nicht zu der Annahme, dass solche Namen im Sinne der Warenzeichen- und Markenschutzgesetzgebung als frei zu betrachten wären und daher von jedermann benutzt werden dürften.

Coverbild: www.ingimage.com

Verlag: LAP LAMBERT Academic Publishing GmbH & Co. KG
Heinrich-Böcking-Str. 6-8, 66121 Saarbrücken, Deutschland
Telefon +49 681 3720-310, Telefax +49 681 3720-3109
Email: info@lap-publishing.com

Approved by: Dublin, Dublin City University, 2011

Herstellung in Deutschland (siehe letzte Seite)
ISBN: 978-3-659-10853-2

Imprint (only for USA, GB)
Bibliographic information published by the Deutsche Nationalbibliothek: The Deutsche Nationalbibliothek lists this publication in the Deutsche Nationalbibliografie; detailed bibliographic data are available in the Internet at http://dnb.d-nb.de.

Any brand names and product names mentioned in this book are subject to trademark, brand or patent protection and are trademarks or registered trademarks of their respective holders. The use of brand names, product names, common names, trade names, product descriptions etc. even without a particular marking in this works is in no way to be construed to mean that such names may be regarded as unrestricted in respect of trademark and brand protection legislation and could thus be used by anyone.

Cover image: www.ingimage.com

Publisher: LAP LAMBERT Academic Publishing GmbH & Co. KG
Heinrich-Böcking-Str. 6-8, 66121 Saarbrücken, Germany
Phone +49 681 3720-310, Fax +49 681 3720-3109
Email: info@lap-publishing.com

Printed in the U.S.A.
Printed in the U.K. by (see last page)
ISBN: 978-3-659-10853-2

Organising for Innovation

A case study of innovation teams
and team leadership in a large,
R&D-intensive, global firm

Peter Robbins, MBS (Research), PhD

Abstract

Successful innovation is vital for firm survival and success (Dougherty, 2004). Leaders of established companies acknowledge that radical innovation, in particular, is critical to their growth and renewal (Leifer et al, 2000). This is especially true for the research and development sector (Eisenbeiß and Boerner, 2010). New product development, while not the only form of innovation, remains the most advanced, the most widely studied, and the most significant type of firm-level innovation (Garcia and Cantalone, 2002). Despite an increase in research on innovation, the identification of specific ways to improve firms' innovation performance, specifically with regard to radical innovation, remains a significant challenge for researchers of innovation (Bessant et al., 2010).

The context for this study is a revelatory case study of two competing R&D teams in a global, high-tech, research-intensive organisation. The teams had a mandate to develop radical innovations, though were characterised by differences in approach and leadership styles. A thematic analysis explores the processes through which each team generated, incubated and, ultimately, implemented new commercial ideas. Four themes emerge from the case study: structure, process, networks and leadership. The analysis suggests that variation in these four elements may explain the variation in the teams' outcomes in terms of radical and incremental innovation.

This study contributes to our understanding of how to organise for innovation, and specifically, how team leadership and networks relate to innovation outcomes. Specifically, it suggests that the three phases of the Innovation Value Chain (Hansen and Birkinshaw, 2007) have differential potency in their likelihood of delivering radical innovation; with a focus on the first phase more likely to produce radical innovation.

Table of Contents

List of Tables

List of Figures

Chapter One

Introduction

1.1   Organising for Radical Innovation

This study focuses on new product development within an R&D department of a large, global, pharmaceutical corporation. New Product Development (NPD) is an issue that is widely accepted as critical for both individual firms and the economy as a whole. Ahmed and Shepherd (2010) note that the potential benefits accruing from innovation make it possibly the single most important activity for organisations.   Specifically, product innovation or new product development (NPD) is widely recognised as an important priority for companies (e.g. Hill and Utterback , 1980; Lundvall, 1985; Cooper and Kleinschmidt,1987; Rommel, 1991; Eisenhardt and Tabrizi, 1995; Dougherty, 1996; Utterback, 1996).

The literature on innovation is diverse and complex, and overlaps with many different topics and research fields (Antonelli, 2009). This complexity makes it challenging for practitioners and scholars to understand and explain the wide-ranging organisational factors that can influence a firm's innovation capability and performance. Several decades of research into innovation management have failed to provide clear and consistent findings or coherent advice for managers (Tidd, 2001). Indeed, managing innovation has been recognised as a long-standing and intractable problem (Salaman and Storey, 2002) mainly, according to Igartua et al., (2010) because the concept comprises so many, diverse component parts that guidelines inevitably become fragmented.

There is, however, agreement in the literature that proficiency at new product development is a key priority for managers (Van de Ven, 1986) and for their organisations (Cooper and Kleinschmidt, 1994; Kessler and Chakrabarti, 1999; Liker et al., 1999; Sethi et al., 2001; Bonner et al., 2002; Filippini et al., 2004; Szymanski et al., 2006). Arguably, Tushman and Nadler's assertion about the prevailing mood about innovation, therefore, still holds relevance:

> In today's business environment, there is no executive task more vital and demanding than the sustained management of innovation and change...to compete in a new environment, companies must create new products, services and processes...they must adopt innovation as a way of corporate life (Tushman and Nadler, 1986: 74)

However, despite this scholarly interest in NPD structures, the question of how firms should implement an effective new NPD process design for decreased cycle time and increased innovation productivity remains largely unanswered. Precise knowledge of the organisational factors that lead to radical innovation, in particular, is underdeveloped to the extent that success is often attributed to serendipity (Bessant et al, 2004). Wolfe asserted that; 'Our understanding of innovative behaviour in organisations remains relatively under-developed' (1994: 405).

Key questions remain in the literature and in the field about just how radical innovation, in particular, emerges within large organisatons (Leifer et al, 2000). This includes specific details like: how R&D teams organise for innovation; how they generate new commercial ideas; how they develop, rank, incubate and prioritise their ideas; how they convert raw ideas into testable concepts; how they research them; how and why teams prioritise and champion certain ideas and abandon others; how teams get ideas implemented and adopted by the organisation as a whole; and how such teams are and ought to be led. These qestions form the key focus of this study.

Companies constantly search for proven, effective methods to innovate. But the quest for organisational factors that are proven to be successful in the innovation process is a significant challenge for many reasons. By its very nature, innovation is largely unpredictable and requires flexibility, opportunism, and adaptability; it's a delicate balancing act (Kanter, 1989). Similarly, there is agreement that due to the complexities associated with innovation research, there will never be just one true theory or best practice of innovation (Tidd, 2001; Thamain, 1990; Harmancioglu et al., 2007). It is these continuing gaps in knowledge about what really happens inside firms which leads to successful innovation outcomes, combined with the fact that this potential to achieve more with innovation remains largely unexploited by management practitioners that provides the motivation underpinning this thesis.

In the research on innovation, what is clear is that different theories apply in various contexts within an organisation or even within individual projects, such as: the stage of development of the innovation (Damanpour and Gopalakrishnan, 2001); the type, scope or ambition of innovation pursued (Christensen, 1997; Leifer et al., 2000); the life-cycle stage of the organisation (Koberg et al., 1996; Sorensen and Stuart, 2000); the level of customer involvement in the process of co-evolution of innovation (Von Hippel, 1986, 1988; Prahalad and Ramaswamy, 2003); the type of leadership assigned to the project group or innovation team (Buijs, 2007); the phase of the innovation value chain (IVC) in which the project is situated (O'Connor and Ayers, 2004; Hansen and Birkenshaw, 2007; Roper et al., 2008); particular characteristics of public service innovation (Damanpour and Schneider, 2009) the corporate priority assigned to the brands or assets involved (Birkinshaw and Robbins, 2010); whether it is a product or service innovation and the role of design (Bessant and Maher, 2009); and the wider environment that the organisation operates within (Koberg et al., 1996; Brennan and Dooley, 2005).

This contingency approach to describing new product and service development within firms is common in innovation research (e.g. Wolfe, 1994; Damanpour, 1996), but what is also important is the role of the relationships between the factors that influence an organisation's ability to manage innovation. The relationships between factors will be influenced by organisational context such as organisational size, age, resources and competitive elements and the external environment. The culture in the organisation; the work climate for creativity and the team climate for innovation; the use of cross functional teams; the engagement with external network partners; the management support for innovation and, crucially, the leadership of the innovation teams will also influence innovation efforts (Phillips et al, 2006; Amabile et al, 1996, 2006; Anderson and West, 1998; Balsano et al, 2008; Barczak and Wilemon, 2003; Caldwell and O'Reilly, 2003; Cohen and Bailey, 1997; Cooper, 2009; Edmondson and Nembhard, 2009; Goffin et al, 2010; Keller, 2006; Kratzer et al, 2006; Mumford et al, 2007) .

The problem of how to manage innovation, however difficult, is not new and the sources of the problem have been investigated at many levels and from diverse perspectives. Most of the studies exploring innovation adopt a positivist perspective (Biemans et al, 2010). That is, they treat innovation as the dependent variable and try to establish, identify and, ultimately, measure the influence of a series of independent variables. In this way, they seek to explain the scope, pattern and degree of innovation within organisations (Salaman and Storey, 2005). This study takes a different approach as it explores variation in innovation outputs and links them directly to the processes employed.

Researchers often either focus on how the process of innovation is organised in firms (innovation modes) or, alternatively, on the results of the innovation processes and their characteristics (innovation outcomes). In the literature on innovation management these two aspects are treated in a somehow separate way (Isari and Pontiggia, 2010).

R&D is often the starting point for innovation and new product development. The R&D team in any firm, which is responsible for most of the organisation's product innovation, plays, probably, *the* crucial role in firm survival (Huang and Lin 2006). R&D projects are generally complex, time-pressured and uncertain. R&D is risky: vast amounts of money can be lost in attempts to innovate (Stevens, 1996) and critics of innovation often suggest that stock markets respond negatively to innovations which have a longer lead time (Sood and Tellis, 2009). Thamhain (2003) argues that, as with innovation generally, no single, blueprint exists that guarantees success in R&D. However, the R&D innovation process is necessarily purposeful. Therefore, the quest for more meaningful insight that might help guide managerial action to more effective performance is important. In this context, Goffin et al (2011), assert that the ability of organisations to learn from prior experience is essential if they are to improve their performance at innovation: 'Sustained improvements in R&D depend on the capacity of an organisation to learn' (p. 301). But although, such learning has been noted as a requirement for the successful development of innovation (Drejer and Riis, 1999; Prahalad and Hamel, 1990); the NPD process has proved a difficult learning environment for R&D organisations (Michael and Palandjian, 2004). Arguably, this is because the new product development process is highly complex, involves cross functional integration with many individuals involved and this gives rise to numerous uncertainties (Brettel et al, 2011). Also, research into creativity in R&D has not been extensive and hence understanding of the area is arguably underdeveloped (Eisenbeiß and Boerner, 2010; Kurtzberg and Amabile, 2001;West, 2002; Anderson et al, 2004).

A flexible organisation, with high levels of freedom and risk taking, is commonly proposed as the ideal springboard for innovation (Phillips et al, 2006; Nohria and Gulati, 1996). However, extant evidence suggests that it is difficult to achieve projects that run efficiently but that still leave room for exploration and the creation of new knowledge (Smith et al., 2008).

In this context, it is also important to make a distinction between two types of R&D that firms undertake: exploration and exploitation (Schumpeter, 1934; March, 1991; Lavie and Rosenkopf, 2006). Firms undertake exploration in R&D to create new products and deploy new technologies. Exploration or exploratory research (including exploration alliances) is characterised by long time horizons and unpredictable, high variance returns (Hoang and Rothaermel, 2009). On the other hand, firms undertake exploitation in R&D to improve product lines within markets that they are currently serving (Hoang and Rothaermel, 2010; Rothaermel and Deeds, 2004). This distinction has parallels in the classification of types of innovation with exploration correlated to radical innovation and exploitation more likely to yield incremental innovation (Nelson and Winter, 1982; Dewar and Dutton, 1986; Leifer et al., 2000). There is increasing evidence that firms are favouring radical innovation because, when successful, it offers superior return on investment (Bessant et al., 2010). Therefore, any insights or guidelines that could enhance firms' capacity to successfully deliver radical innovation will have considerable managerial relevance.

Any NPD process requires a high level of creative performance. In innovation, creative performance is generally mediated through a team or working group rather than an individual. According to Leenders et al, (2007), creative performance is of paramount importance in NPD projects and most NPD projects are managed through an NPD or innovation team as the organisational nucleus for innovation. The value of teams in new product development is well established. Both the complexity of the work; the blend of skills and expertise required and extant best practice suggest that when professionals from different functions work together on development projects they create the most successful product in the shortest possible time (Edmondson and Nembhardt, 2009). As creative and innovative work, within the R&D setting, is complex, it demands the collaborative efforts of creative people with different areas of expertise (Mumford et al., 2002). Such complexity makes successful performance of innovation activities especially difficult and puts additional strain on leaders of innovation teams.

Research surveys have been conducted to determine the organisational factors in firms that are most predictive of success in NPD initiatives (Cooper and Kleinschmidt, 1999; Cooper et al., 2008; Cooper, 2009; Cooper and Edgett, 2010; Barczac et al., 2008). For instance, the Product Development and Management Association (PDMA) recognises the importance of new products to organisations and has undertaken three major studies on "best practices" in NPD (Barczak et al, 2003; Griffin, 1997; Page, 1993). Overall, these, and other (Adams, 2004), large scale studies focus on the number and types of activities conducted by organisations during the NPD process and the relationships between the proficiency in executing these activities and new product performance. Such surveys generally identify issues like having an explicit innovation strategy; having an effective innovation process (Adams, 2004; Griffin, 1997); having cross functional teams managing the process; they often refer to the quality of the people on the innovation team and usually specify that the best companies ensure that there is senior management responsibility for the results of the innovation programme. Bringing the voice of the consumer (or customer) into the NPD process has generally been a feature of best practice in NPD. However other issues such as, the selection of team leader; the leader's characteristics, experience; management style; their networks both inside and outside the company; their skills; their technical knowledge and their level within the organisation have rarely feature in these frameworks.

Strong team leadership, for innovation and R&D teams, includes the ability to direct and coordinate the activities of other team members, assess team performance, assign tasks, develop team knowledge, skills, and abilities, motivate team members, plan and organise, and establish a positive atmosphere (Tannenbaum et al., 1998; Eisenbeiss et al., 2008; Messinger, 2009; Zheng et al., 2010). As boundaries between the team and the stakeholders will be porous, strong team leadership requires being clear about which are the unhelpful information flows across team boundaries during creative phases (pressure, delivery dates, high expectations and threats) and which are the helpful flows (information, contacts, experience and alternative solutions). Cooper and Edgett (2008) argue that leadership in an innovative context will require vision and keeping situations open; being flexible, experimental and positively disposed to creativity rather than focussing on the solution and shutting situations down. Kolb et al, suggest that an effective leader for innovation will manage team boundaries with the external world and let their teams concentrate on their objectives (1993). However, academic examination of leadership influence on R&D team innovation has been, at best, inadequate and often controversial (Bass, 1999; Keller, 1999 and 2004; Nippa, 2006; Stoker et al., 2001).

That leadership is important in managing innovation teams is established (Senge, 2006; Van Buijs, 2007; Sarin and O'Connor, 2009). Similarly, there is evidence that innovation projects, in getting from raw idea to its eventual introduction or implementation, follow a journey with very distinct stages, namely: Idea Generation, Idea Conversion and Idea Diffusion. (Cooper and

Kleinschmidt, 1997; O'Connor and Ayers, 2004; Hansen and Birkinshaw, 2007; Roper et al., 2008). It is arguable that different types of team leadership will have differential potency in an innovation project depending on the stage of progress of the individual project.

In summary, innovation and, specifically, proficiency at new product development is a vital element of most firms' ability to survive and compete. The majority of firms who are innovation-active rely on cross function teams as the organizing unit through which innovation is delivered.

Radical innovation is associated with organisations with the capacity for continuous learning (Bessant and Francis, 1997; McLaughlin et al, 2008) and teams , 'rather than individuals, are the fundamental learning unit in most organisations' (Senge, 2006; p.10).

R&D is the start point for most innovation projects, in large firms, and R&D shoulders the burden for delivering the majority of companies' innovation output. But, the operating context for innovation is characterised by risk and most projects fail (Christensen, 1997; Foster and Kaplan, 2001). This study seeks to identify some of the important factors which both encourage and frustrate the successful development of new product and service ideas within innovation teams. It examines the structure, processes, networks and leadership of an innovation project in a large organisation to determine their impact on the outcomes. Such an understanding is not only critical for organisations but also aligns with Ireland's national ambitions of competing on the basis of creativity and innovation.

1.2 The Importance of Innovation in the Irish Context

Recent reports from the Irish Government repeatedly assert that innovation and creativity are key determinants of success for Ireland. In particular, a number of recent government reports allude to a vision of creating Ireland's 'smart economy'. The basic idea, as expressed in five reports reports – the 'Strategy for Science, Technology and Innovation 2006-2013' (2006); 'Building Ireland's Smart Economy' published in 2008; 'Sharing our Future: Ireland 2025' (2009); 'Trading and Investing in a Smart Economy; Action Plan to 2015' (2010) and the 'Report of the Innovation Taskforce' (2010) – is to focus public investment on a set of targeted priority areas in science, engineering and technology. The intention is to create a research, innovation and commercialisation environment that will translate knowledge creation into economic activity and ultimately generate highly paid and sustainable jobs in Ireland.

Innovation is currently the central theme in the Irish Government's plans for economic recovery. The first formal declaration of this national focus on innovation, 'Building Ireland's Smart Economy' was published by the Department of the Taoiseach (Prime Minister) in 2008. Intended as a blueprint for sustainable development, this programme outlined the Government's commitment to a set of actions to reorganise the economy over the period 2008-2013 in order to secure the economic future for Ireland.

The Smart Economy combines the successful elements of the enterprise economy and the innovation or 'ideas' economy.

> A key feature of this approach is building the innovation or 'ideas' component of the economy through the utilisation of human capital - the knowledge, skills and creativity of people - and its ability and effectiveness in translating ideas into valuable processes, products and services. (Building Ireland's Smart Economy: A Framework for Sustainable Economic Renewal, 2008; p. 7).

A number of subsequent Government Reports have also put innovation at the top of the Irish Government's business and economic agenda. Forfas' (Ireland's state agency responsible for developing industrial policy) report 'Sharing Our Future' (2009) stated:

> Innovation in all its dimensions will continue as *the* central driver of wealth creation, economic progress and prosperity in the coming decades. Innovation will no longer be about technological innovation but will include organisation and business model innovation, workplace innovation, creativity and design. (Forfás: Sharing our Future: Ireland 2025; 2009; p. 6. Emphasis added).

In order to frame specific policies to facilitate Ireland's transition to a smart economy, an *Innovation Taskforce* was appointed by the Government in July, 2009. This group, which included some private sector representation, was mainly comprised of state agencies. It included the heads of the state development agencies, the Industrial Development Authority and Enterprise Ireland as well as the heads of Ireland's two largest universities (Trinity College and UCD). In 2010, the Report of the Innovation Taskforce states:

> What we need to do now is to place innovation at the heart of enterprise policy. Our future economic success depends on increasing levels of innovation across all aspects of Irish enterprise – from large Irish-owned multinationals to foreign multinationals located here to established Small and Medium Enterprises (SMEs) in services and manufacturing, as well as start-ups and existing companies with high growth potential. (Report of the Innovation Taskforce, 2010; p. 2)

Within this report, the Government unveiled its ambition to create 'an Innovation Island' and to make Ireland an innovation hub for Europe:

> The key objective of this *Action Area* is to make Ireland an innovation and commercialisation hub of Europe – a country that combines the features of an attractive home for innovative multinationals while also

being a highly-attractive incubation environment for the best entrepreneurs from Ireland and overseas. (Building Ireland's Smart Economy: A Framework for Sustainable Economic Renewal, 2008; p. 13)

That specific policy intervention, to help Ireland succeed with innovation, is warranted has been repeatedly demonstrated by Ireland's performance in the EU Community Innovation Survey (CIS). The Community Innovation Survey (CIS) has been conducted a total of six times; 1992, 1996, 2001, 2002-4, 2004-6 and 2006-8. The CIS defines a firm as being innovative, or innovation-active, if it introduces at least one product, service or process that is new to the firm itself within the period under review (Arundel et al, 2007).

The Irish Community Innovation Survey for 2008-2010 collected information about product and process innovation as well as organisational innovations and other key variables. Most questions covered new or significantly improved goods or services or the implementation of new or significantly improved processes, logistics or distribution methods. Compared to other EU member states, Ireland continues to be seen as a 'follower' in innovation. The CIS reports a general trend suggesting that the larger the firm, the more likely it is that it will be engaged in some mode of innovation. It notes that Irish firms are generally less innovative than their foreign-owned counterparts. Another finding is that when smaller firms innovate, they are much less likely than larger firms to engage in any collaboration with external partners.

In examining barriers to innovation, the survey asks why companies are not engaging more (or at all) with innovation. They conclude that the principal barrier is funding. Companies see innovation as expensive and have difficulty in getting the funds to innovate. Second only to access to funds to innovate is the issue of qualified personnel; this is followed by lack of information on technology. Firms report that they do not have the expertise or specialist personnel they would wish to have in order to voyage into the area of new product and service development. This suggests that there is a lack of process know-how within Irish firms about how innovation really unfolds in organisations and what might constitute best practice in the area.

What can be concluded from the Irish innovation context is that innovation (new product, service, and business development) is high on the Government's agenda and that it is likely to remain a high national priority. Innovation is a top priority for business and therefore any practical insights that may offer a better framework for enhancing innovation performance will be a welcome addition to both policy and practice. An objective of this thesis is to investigate; how innovation really happens in firms and what sort of practices organisations should embed in order to lead to successful new product and service development. Hence, this thesis is conceptually relevant both for practitioners and policymakers.

1.3   The Case Research

This thesis is based on a 'revelatory' case study of what could be described as an organisational experiment in innovation in a large R&D intensive firm. The case study is based on two teams that 'competed' to develop 'radical, game-changing' new product ideas for the firm. This case study represents an interesting context for several reasons. First, the firm, GlaxoSmithKline (GSK), was the world's second largest pharmaceutical company with large R&D expenditure (GSK was one of the world's top ten spenders on R&D in 2009). Second, as part of an effort to develop 'bigger and better' ideas to fill its new product development pipeline GSK created a project called *Innovation Sans Frontiers* (ISF). This project involved two 'Innovation Teams' working to the same brief but in competition with each other; it was time bound, a nine-month project to end in the summer of 2007; and the outcomes were 'judged' by senior management, with one team 'winning the contest'. Third, a priori, the project was designed to be 'different' from existing approaches to innovation within GSK. For example, the project was designed to provide additional management autonomy for the participants; the teams were encouraged to explore areas in which they had a strong personal interest, including opportunities outside the existing sectors and therapy areas in which the firm operates; and the team leaders were not required to account for their time or expenditures. Fourth, the two teams were comprised of roughly 12 members and were 'created equal' insofar as was possible with their collective experience, expertise and seniority balanced across the two teams. Finally, the two teams were based in different geographic contexts (the UK and the US).

The study will show how one team produced innovation ideas and concepts (outcomes), which could be classified as radical while the other team's outcomes were described as less original and more incremental in nature. In this study I was given the type of access to the people and data surrounding this project which is rarely forthcoming from global organisations. As such, the study provides a unique opportunity to evaluate the managerial practices currently deployed with the objective of identifying the elements of those practices which should be encouraged and those whose value is more equivocal. Thus, it offers an opportunity to respond to the questions in the innovation literature.

1.4   The Research Objective & Research Question

In doing this study my objective was to undertake an in-depth and longitudinal examination of an innovation project or programme within a large, global, R&D intensive organisation (GSK) to gain a deep, rich understanding about the innovation process, which may have relevance for developing and advancing the practice of new product development and innovation management more generally. Specifically, I sought to review the outputs of the innovation contest and to assess the influence of various antecedent factors in arriving at the outputs.

This study explores how innovation unfolds within teams, who themselves operate within organisations. It focuses on how radical innovation can be encouraged within large organisatons. It explores issues like: how teams actually create and develop new ideas; how they sift, rank, incubate and prioritise them; how they convert raw ideas into testable concepts; how and why teams or individuals champion certain ideas and abandon others; how teams get ideas implemented and adopted by the organisation as a whole; how they find sponsors in senior management to facilitate the assimilation of their ideas into the company's pipeline. It also examines how such teams are and ought to be led. This latter element, leadership of R&D teams, is important to this study as the case facilitated an analysis of the type of leadership most appropriate for each phase of the innovation process, value chain or 'journey'.

1.5 The Research Methodology

1.5.1 What was Done?

This study is based on a single, revalatory case-study. The approach is inductive. The study was constructed in a series of three consecutive phases to provide increasing focus to the investigation.

The three phases are illustrated in Figure 1.1 (below). The first phase involved interviewing senior management in GSK's worldwide HQ (Generally at functional 'President' or 'Vice President' level) to identify the most appropriate case within GSK's extensive, global R&D operations to study. The factors that led to the selection of the ISF programme included a desire to examine a dedicated, discrete, time-bound programme in order to be able to elicit some meaningful insights about the management of innovation in future initiatives of this type. A further consideration was that this project was explicitly mandated to search for radical ideas. The first phase of the fieldwork involved interviews with the global heads of Innovation, R&D, Organisational Development,

Marketing Excellence and Global Brand Marketing. Through these interviews, the ISF project was selected and agreed at a high level within the organisation as a suitable case-study.

The second phase constitutes the principal part of the study. It involved a series of in depth, semi-structured, mainly (but not exclusively) telephone interviews with the participants on the ISF programme. These interviews took place over a period of two years. Interviews were carried out with everyone who was a member of either team in the initiative, over the two sites in the UK and US with the exceptions of a couple of individuals who had left the organisation either during or fairly soon after the conclusion of the project. Along with the interviews, there was additional material ranging from briefing slides, project plans (US only), the two teams' PowerPoint slides and videos for the final presentation of their outputs in New Jersey, internal close-out reports on the project along with considerable consumer research data and video material.

The third phase involved interviews with the judging panel for the innovation competition. Members of this group sit on GSK's Senior Leadership Team (SLT) within R&D and are responsible for the number, scope and progress of initiatives that are undertaken by R&D. They allocate the budgets and resources and manage the innovation process from the R&D side and they determine the portfolio of projects that are supported, making decisions on the balance between radical and incremental projects. This group are the most influential gatekeepers within the innovation process in R&D.

The three-phased approach allowed for a chronological analysis of the case. The preliminary interviews showed why such an experimental approach was deemed appropriate by the SLT for this project. The participant interviews revealed which were the elements of the project design and management that encouraged innovation and which ones frustrated it. Finally, the interviews with the SLT gave a much richer insight into how senior management expected the initiative to be run; how they ranked the ideas; which ones have survived and are still being worked on in the organisation. These interviews also revealed some differences of opinion within SLT as to which group should have won the contest as it transpired that the 'declared winner' was not the universal choice.

1.5.2 How it was done?

Figure 1.1: The Research Process for this Study

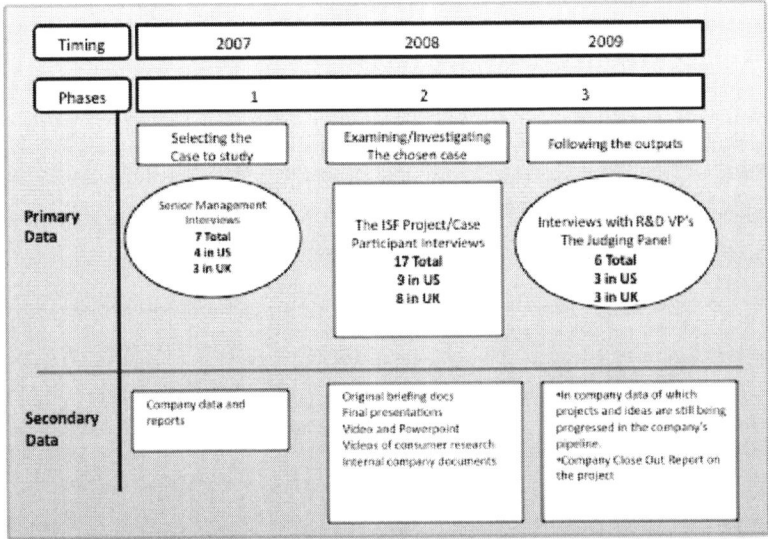

ISF Case Study research programme PR DCU

Source: Case Study Records

1.6   Findings

This study's analysis identified (i) the factors that encouraged or inhibited the participants from creating and developing ideas for new products, and (ii) the influence of internal and external networks on the type and novelty of the ideas and (iii) the effect of team leadership on the innovation process.

The factors that enabled or inhibited participants to develop ideas reflect many existing themes in the literature. Specifically, the study suggests four key areas in developing a firm or project-level infrastructure conducive to innovation: structures, processes, networks and leadership.

This study has successfully identified a number of team or project-level factors, which seem to enhance organisations innovation performance. High among these is the issue of leadership for innovation teams; for innovation projects and, specifically, for the phases of individual innovation projects. Additional considerations which are predictive of positive NPD outcomes surround the initiating structures of innovation teams (the time allocated for innovation and, specifically, time devoted to each section of the innovation value chain; creative spaces for innovation; squaring off line managers; the size of the team itself) and the processes (integrating the voice of the consumer; converting raw ideas into testable concepts and effectively screening the ideas; and finding a way to prioritise ideas which is genuinely objective) they use. A further critical consideration is the use of internal and external networks.

In terms of team leadership, the case study revealed that the leaders of the two teams adopted very different approaches to the task. In summary, the UK team leader abandoned any formal team processes and allowed team members to work on their own. Additionally, the UK team leader was a scientist, of some repute in his field, and he brokered some important introductions to other specialists and networks outside the organisation.

The US team, in contrast, was led by a former project manager, who adopted a structured approach to the work of the team. During the project he 'sensitised' senior management to the nature and types of ideas on which his team was working. At the end of the competition the US team was declared the winner, although follow-on discussions with senior management suggests that many of the (losing) UK team's ideas may have a longer lasting impact on the development of new products and services at GSK. Moreover, the majority of the judging panel considered the ideas and concepts from the UK to have been far more imaginative and radical than those of the US team.

The approaches adopted by the two team leaders in managing their teams have been classified in terms of transformational and transactional leaders (Bass, 1985). The team leader from the UK could be described as a transformational leader, while the US counterpart was more transactional, adopting a structured approach to the task.

Activities across the project covered a spectrum of 'exploration' and 'exploitation' and the case study suggests that different forms of leadership have a differential impact depending on the stage of the innovation process. It was further identified that the UK team delivered a far higher number of ideas which were classified as potentially radical ideas for the organisation while the ideas put forward by the US company, while well-researched, were seen as largely incremental.

The study suggests that transformational leadership is more appropriate for teams at the 'R' stage. In contrast we find the 'D' stage, where the focus has shifted to exploitation, transactional leadership is more appropriate. As such we argue that neither leadership approach alone represents an 'ideal' approach to spanning the full spectrum of the innovation process (Hansen and Brkinshaw, 2007) from creativity to implementation.

1.7   Structure of the Thesis

The thesis is structured around an in-depth case study from within GSK. There are nine chapters in the thesis for which the titles and objectives are described below. Chapter one sets the context for the study by identifying innovation as a key priority for industry and for policymakers. It notes the concern of the Irish government that many Irish businesses are not performing well in terms of their innovation outputs. It also describes the research objectives, the research methodology and design of the study.

Chapter two contains a review of the literature around innovation and new product development. It establishes the theoretical framework for the study of innovation management. It provides a definition of key terms; a description of frameworks and identifies how innovation has been generally measured and researched in academic studies. This section identifies some gaps in the literature around what is known about how the innovation process happens in large complex organisations.

Chapter three outlines the research philosophy underpinning this study. It describes, in detail, the research design and the research process.

Chapter four, by way of general context, provides an overview of the pharma sector in 2007 (the year in which the case is set). It describes key global trends in the industry and reviews their implications for some of the leading pharma companies. Specifically, it concentrates on GlaxoSmithKline (GSK) and examines its Consumer Healthcare (GSK CH) business division. The chapter describes GSK's approach to new product development and situates the Innovation Sans Frontiers (ISF) project within the organisation's overall innovation programme within R&D.

Chapter five describes the background and objectives for the ISF project. This chapter contains the fieldwork for the study and it tells the story of the ISF initative from the perspective of the participants. It contains chronoligical, verbatim reports of the team members as they describe their experiences with the teams. It charts the progress of both teams and concludes with interview excerpts from the R&D Senior Leadership Team (SLT) who acted as the judging panel for the teams's outputs.

Chapter six presents a thematic analysis of the field data. It identifies, through axial coding, the issues that emerged as the most significant from the fieldwork and categorises these into key operational and managerial issues that are likely to either encourage or inhibit innovation.

Chapter seven looks beyond the participant experiences to provide a theory-based context for this study. It analyses the case looking at the prject outcomes rather than the team or individuals as the unit of analysis. It also attempts to map the project across the innovation value chain. In its review of the project outcomes, this section draws conclusions about the number, type and quality of the ideas or concepts developed by each team, specifically, classifying the outputs on a spectrum of incremental to radical.

Chapter eight reviews the relevant literature to situate the key elements of the case study in a discussion with the appropriate theory. Within this chapter, a theoretical model is developed which fuses the practice-based insights, generated by the case, with the relevant theory. The objective of the model is to improve future guidelines for managing innovation projects in GSK and, possibly, elsewhere.

Chapter Nine makes clear the overall contribution this study makes to extant literature in this area and it positions this research in terms of its relevance and import to policy and practice. It concludes by making recommendations for future research.

Chapter Two
Literature Review

2.1   Introduction

The focus of this inductive study was to examine how innovation happens in large organisations.   Specifically, access was negotiated to a novel organisational experiment within the case firm.   The nature of the innovation project was that the organisation was seeking to develop radical innovation through two separate and competing R&D teams.   Prior to engaging with the case, a review of existing literature relating to the core aspects of firm and team level innovation was conducted.   This review is organised around the following three questions: How do firms innovate? Why does managing the innovation process pose such a formidable challenge for organisations?   And, what is the role of teams within the innovation process?

The general management literature often prescribes that organisations need to enhance their organisational innovativeness to remain competitive (Porter, 1990; Lengnick-Hall, 1992; Roberts, 1998), but, as Smith et al. (2008) note, the literature often neglects to address exactly how organisations can impact on their ability to manage innovation. Similarly, Grönlund et al. (2010; p. 106) note that the academic debate has now moved from 'why to innovate to how to innovate'.   As Goffin and Mitchell put it: 'Innovation: yes, but how?' (2010, p1). Research on the innovation process has become both more prevalent and more important as organisations search for new and more creative ways to remain competitive in their respective markets (Pearce and Ensley, 2004; Janssen et al, 2004). The centrality of innovation to the success of any firm is generally accepted (Ancona and Caldwell, 1992). But, despite the scholarly investigation and interest in New Product Development structures and processes (Cooper and Kleinschmidt, 1994; Kessler and Chakrabarti, 1999; Liker and Collins 1999;

Bonner et al., 2002; Filippini et al., 2004; Troy et al., 2001) many questions about how innovation actually takes place; between people within teams that are themselves, within organisations, remain unanswered. Van de Ven and Poole (1990) were asking similar questions 20 years ago when they noted that managers of innovation projects 'need a "road map" that indicates how and why the innovation journey unfolds, and what paths are likely to lead to success or failure.' Arguably, no single, unifying guideline exists that guarantees success in R&D (Tidd, 2001; Thamain, 2003; Harmancioglu et al., 2007). Therefore, the quest for more meaningful insight that might help guide managerial action to more effective performance is important. Innovation, to succeed, requires a complex social interaction between both internal and external players (Kuandykov and Sokolov, 2010) and is highly context dependent (Hansen and Wakonen, 1997).

Considerable debate surrounds the definition and boundaries of the term 'innovation' (Adams et al, 2006, 2011; Trott, 2005; Tidd et al, 1997). In its broadest sense, this phrase refers to the invention and implementation of a novel idea, relative to a social context, with the purpose of delivering benefit. Tidd et al. (1997) offer a definition that summarises this breadth:

> ...innovation is often confused with invention but the latter is only the first step in a long process of bringing a good idea to widespread and effective use (p. 24).

Drucker saw innovation and entrepreneurship as inextricably linked (1986):

> Entrepreneurs innovate. Innovation is the specific instrument of entrepreneurship. It is the act that endows resources with a new capacity to create wealth. (p.30)

In this study, the Tidd and Bessant (2009) definition is adopted which describes innovation as 'the process of turning opportunity into new ideas and of putting these into widely used practice' (p16). Although all types of innovation are important for companies, the case studied in this thesis is focused on product innovation or new product development (NPD). Many researchers have identified new product development as the most important form of innovation (e.g., Rommel, 1991; Dougherty, 1992; Janz et al., 2001; Hauser et al, 2006). Sandberg (1992) states that Schumpeter's "view of innovation is consistent with the current focus on product innovation". Similarly Pleschak et al. (1994) say "product innovation is the most important element " and Pleschak and Sabisch (1996) point out that product innovation is one of the most important sources of profit for firms.

Despite increased academic research, the processes by which innovative ideas are generated, incubated and implemented within large, R&D intensive companies, are still 'unclear' (Sundstrom and Zika-Viktorrson, 2009). Many articles and surveys address innovation at the level of the country (Pianta et al., 2008; Puga et al., 2010) or region but innovation, in the form of developing new products and services, actually happens at the level of the (R&D) team or the project within the firm. Despite this, the amount of research dealing with R&D team innovation is rather small (Kurtzberg & Amabile, 2001; West, 2002; Anderson et al, 2004). As Anderson and West (1998) noted,

> Comparatively few studies have focused at the level-of-analysis of the work group. This is a notable shortcoming because it is often the case that an innovation is originated and subsequently developed by a team into routinised practice within organisations. (p. 239).

Studies looking at firm-level innovation identify a range of factors likely to contribute to success in the new product or service development. These generally include having a specific strategy; having robust processes; using a cross functional team; having both technical knowledge and marketing insight (Barczak et al., 2008); but one factor that invariably emerges as significant, specifically in the field of innovation within R&D, is the quality of team leadership. Yet, despite it being identified as a critical component (Brown and Eisenhardt, 1995), many authors argue that this facet of new product development has either not been adequately investigated (Mumford et al., 2002) or such investigation has been inadequate and controversial (Bass 1999; Keller 2006; Nippa 2006; Stoker et al., 2001; Yukl 1989). Nippa (2006), in relation to team leadership for innovation projects, suggests:

> Not surprisingly, findings, insights, and recommendations remain ambiguous and appear to be a patchwork of loosely coupled insights rather than the product of a progressive research agenda. (p. 1)

Buijs (2007) argues that: 'The issues of staffing the innovation team and selecting the people who are going to lead the innovation process have hardly been discussed in the innovation literature.' (p. 203)

However, in this review, it is argued that Nippa's (2006) assessment and Buijs' (2007) contention are not entirely complete as there is, indeed, a literature that explores R&D leadership. While R&D and innovation are not fully synonymous; R&D is the central engine, core activity and starting point for innovation in many large firms (Kratzer et al., 2006); a fact which makes R&D and innovation, at a minimum, overlapping. Hence research into successful R&D team leaders has considerable relevance when looking at leadership in innovation teams.

As the research methodology was inductive and longitudinal, the majority of the reading and synthesis occurred during the research activities themselves, thus some literature will also be introduced in later chapters (Chapter Eight) along with the appropriate results and discussion. This chapter will conclude with some current debates in the literature, focusing on the key questions related to this study. It will use the literature to develop more substantive questions related to the central issue.

2.2 How Firms Innovate?

2.2.1 Models of Innovation

When looking at innovation at the level of the firm, 'there are two dominant theories or perspectives on innovation and new product development' (Conway and Stewart, 2009, p. 65). They are broad and abstract in nature but play a key role in helping the conceptualisation of the contextual relationships inherent in innovation projects; between the organisation, the marketplace and the project itself. This section introduces these two broad models and discusses them as a prelude to evaluating the featured ISF case study from the perspective of both. These two perspectives, a linear perspective (Pinch and Bijker, 1989; Akrich et al, 2002; McCarthy et al, 2006) and a network perspective (Tsai, 2001; Akrich et al., 2002; Ritter, 2003; Leenders and Van Engelen, 2003; Chesbrough, 2003; Dechow and Mouritsen, 2005), offer two very different accounts of how innovation happens in organisations and how innovation activities should be managed.

2.2.2 The Linear Perspective on Innovation

The linear perspective on management of innovation assumes a project-management method and process that delivers appropriate outputs at specified junctures and within predetermined cost limits. The linear perspective characterises innovation as a series of individual, discrete and sequential steps that can be managed by senior managers using analytical techniques. Gate models (Griffin, 1997) are the most widely used frameworks to manage innovation processes. Each phase involves a specific activity and concludes with a decision point (Johnson and Jones, 1957) with managers making informed choices based on codified knowledge of the projects; their appeal to customers, their requirements for technology, their position within a competitor framework and their likely value to the overall company portfolio.

Rothwell (1983) distinguishes two principal types of linear innovation processes. Both the 'technology push' and 'demand-pull' models (See Figure 2.1) conceptualise innovation as a linear, sequential process. The assumption underlying the technology push model is that innovation will emerge from a focus on technological developments and that technology and science will dictate the direction and pace of innovation. Rothwell refers to technology push as a 'first generation' model as it is representative of practice in the 1950's (1983). He describes this as a period of rapid economic growth and expansion which saw the emergence of new sectors such as the emergence of semi-conductors, pharmaceuticals and composite materials driven by technological advances. Ahmed and Shepherd (2010) suggests that under this approach, firms believed that simply doing more R&D would result in more successful products and services and that insufficient attention was paid to the role of the marketplace.

Figure 2.1: Linear Models of Innovation – The 'Science-Push' Model (Top) and the "Demand-Pull' Model (Bottom)

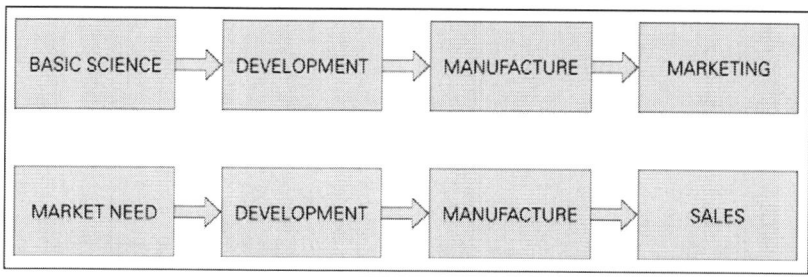

Adapted from Rothwell, 1983

Research in the 1960's and 1970's suggested that most new products (between two thirds and three quarters) could be described as *demand-led* (Utterback, 1971; Rothwell et al., 1974). The increasing importance of users in the innovation process led to the description of "democratization of innovation" (Von Hippel 2005). Rothwell (1983) refers to the demand or market pull models as 'second generation' models which favoured the demand rather than the supply side of the equation. As the descriptor implies, *demand-led* indicates that the impetus for the new product or service came from the customer as opposed to the technology. He characterises this period as one which saw the dominance of marketing, the intensification of competition and the exploitation of technology with greater emphasis on consumer insight and gaining market share. Ahmed and Shepherd (2010) note that this customer centric model was evolved to overcome technology push blindness to the customer requirements of the marketplace. In this demand-led model, the customer is the primary source of the ideas that then drive the R&D programmes. Although, they warn that there are dangers inherent in this approach too:

> One of the primary dangers of following this model is that it produces a tendency to neglect long-term R&D programmes. Thus companies can easily become locked into a regime of technological incrementalism. (p.169)

Figure 2.2. shows an idealised version which merges these two models and is described by Rothwell (1992) as the 'coupling' or 'interactive' model which represents a move away from the definitive linearity of either the push or the pull model. It represents a fusion of customer or market led innovation with the technology push approach. It is still, at heart, a linear model with feedback loops.

Figure 2.2: The 'Interactive' or 'Coupling' Model of Innovation

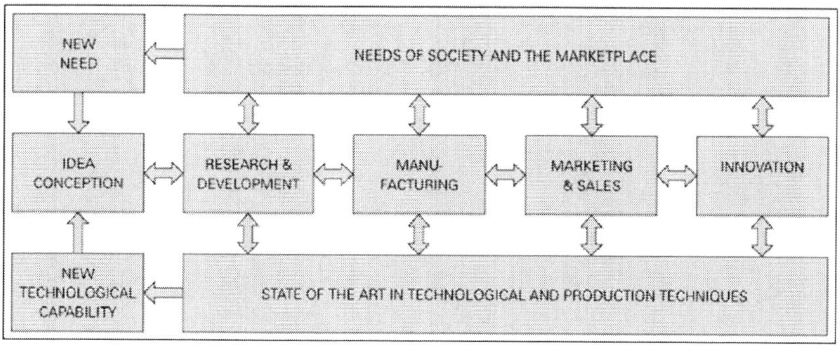

Adapted from Rothwell, 1983.

Rothwell and Zegveld (1984) note that the linkages between the marketplace, technology and science are complex and multidirectional and the dominant element of the equation (the technology or the customer) can vary over time and between industries. Nevertheless, researchers now agree that both demand and supply factors play an important role in innovation and the life cycles of technology (Mowery and Rosenberg 1989; Walsh 1989; Scherer et al, 2000; Lucas 1967; Ben- Zion and Ruttan 1978; Ruttan 1997, 2001).

2.2.3 Design-led Innovation

In more recent literature, a third approach to innovation has been developed which reflects the design-led practices adopted by successful Italian manufacturers (Verganti, 2008, 2009). Design thinking as an approach to innovation argues that not all innovation can be classified as either technology-push or market-pull. This newer perspective on the innovation process that is gaining popularity is based on design-thinking (Martin, 2009). Design-thinking innovation is an approach to innovation that elevates the intrinsic socio-cultural meaning within the products and services. Nominally, based upon the original Latin origin of the word design; 'designare' to give meaning to or to assign meaning; the principle is that the qualities of the new product or service extend considerably beyond merely functional characteristics to also provide enhanced design cues that reinforce their socio-cultural meaning. Dell'era and Verganti (2009) argue that the principles of design thinking are not merely considerations around physical product design and styling:

> A product can bring messages to the market in several ways and styling is just one of them; while the functionalities of a product aim to satisfy the operative needs of the customer, its product meanings aim to satisfy the emotional and socio-cultural needs of the customer. (p 39)

Dell'era and Verganti (2009) cite examples like Allessi and Archimedes (well known Italian lifestyle brands) who use design thinking to add attractive additional dimensions to their innovation ideas and outputs. This approach is also favoured by innovation consultancies such as IDEO (Brown, 2008). Implicit in the design thinking approach to innovation is a reliance on ethnographic research to ensure the ideas are authentically user-centred. Rapid prototyping and customer immersion and involvement in the co-creation of the product or service is also a feature of this approach.

Brown (2008) defines design thinking as a discipline that uses the designer's sensibility and practices to match people's needs with what is technologically feasible and what, through a viable business strategy, can be converted into a valuable market opportunity. Brown (2008) notes that historically designers would have played merely a supporting. 'downstream' role in the innovation process; 'merely to put a beautiful wrapper on the idea' (p. 86). Now, however, the role, the thinking and the methods of designers are being elevated from the merely tactical to be strategic and central to the innovation process.

2.2.4 Structuring the Innovation Process in Stages

Govindarajan and Trimble (2010) argue that organisations today are only marginally more prepared for the challenges of innovation than they were fifty years ago. While most companies, they say, have no shortage of creativity or technology, what they lack are the managerial skills to convert ideas into reality. They liken innovation to an ascent of a formidable mountain peak. In their metaphor, most climbers focus their energy and enthusiasm on getting to the top, leaving very few resources for the less glamorous but often more dangerous part of the expedition—getting back down safely. Similarly, they say, companies devote their energies only to reaching the innovation summit—that is, identifying, developing, and committing to a sparkling and promising idea. In short, they argue, there is too much emphasis on ideas and not nearly enough on execution.

Reflecting the priority of the implementation side of the innovation process and in order to bring more control, structure and standardisation to it, many companies employ a process of milestones or gates. Each gate marks a formal review of predesignated metrics around the project itself. This approach represents 'the overlay of linear thinking on the innovation process' (Tollin and Caru, 2008; p. 75). The gate approach reputedly originated from the National Aeronautics and Space Agency (NASA) in the USA. Their *phased project planning* (PPP) approach was a planning tool to manage the complex co-operations between NASA, its contractors and suppliers (Cooper, 1990; Eppinger, 2001). The PPP divided projects into activities with review points ensuring a high level of control and measurement. It focussed on the technical, design side of projects and did not include issues like customers or marketing. The modern gate approach assumes not only a review of a specific project but it envisages a review of a portfolio of projects , thus ideally choosing the most promising projects to proceed with; the ones which fit best with the overall company strategy (Cooper, 1994, 2001; Christiansen and Varnes, 2007).

The stage-gate approach attempts to control and manage innovation using distinct mechanisms. First, by making resource decisions gradually as more information becomes available, it reduces the risk connected with projects (Christiansen and Varnes, 2007). This implies that projects will be either progressed or abandoned depending on the likelihood of their commercial success. Hurdles (which generally include a number of pre-agreed project performance criteria) are adopted by the company and only if the project meets these criteria is it allowed to proceed. This method also assumes that the outcome of any specific, individual innovation programme is continually in doubt and further work is only sanctioned incrementally as each piece of new information is processed by the decision making team on the project. Also, this approach has the effect of making certain activities mandatory within every project. Conforming to defined stages, activities and hurdles means that all projects must meet a common standard and provides the safeguard that individual projects cannot get green-lighted or progressed without meeting the corporate criteria.

Cooper (1988) coined the term *Stage-Gate* to describe this process. The Stage-Gate process consists of a series of stages where essential activities are carried out. The stages are activities which culminate at gates where interim achievements are evaluated. It is the stages that entail the actual development work. The specific activities performed depend on which stage the project is in. As Figure 2.3 indicates, in the early stages, activities generally focus on discovering opportunities and generating ideas, while the later stages concentrate on concept development, testing, and commercialisation. Stages are typically cross-functional and each activity is undertaken in parallel with others so as to enhance speed to market. Each stage typically requires more investment than the preceding one, resulting in increased commitments but also in a reduced number of unknowns and uncertainties so that risk is effectively managed (Cooper, 1988, 1994, 2008).

Figure 2.3: The Stage Gate Model of Innovation

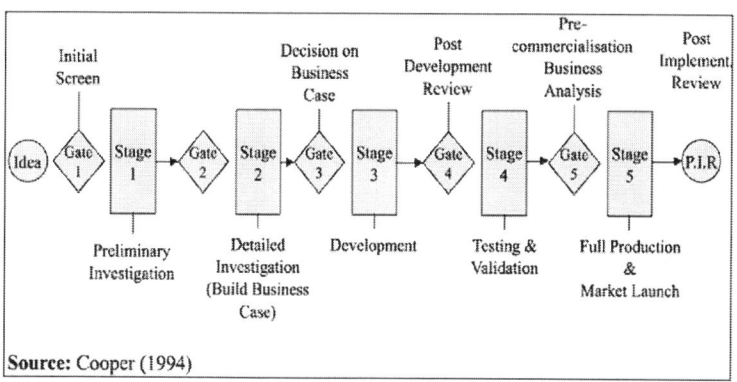

Source: Cooper (1994)

The model above describes the phases in a generic stage-gate model. However, companies tend to customise these models to suit the specifics of their own business and versions with up to ten stages have been reported (Philips et al.; 1999).

The five phases (depicted above) span the entire innovation spectrum from ideation (idea-generation) to launch activities and subsequent post-launch evaluation. This structured, linear model for managing innovation has become widespread in nearly every industry with the possible exception of services. After surveying NPD best practices, Griffin (1997) noted that 60% of responding NPD functions were using some form of Stage-Gate methodology. Previous research shows about half (48.6%) of respondents said their companies used a traditional Stage-Gates process, 20% said they had no formal or an informal stage-gate process, and nearly 30% of respondents said they used a modified Stage-Gates process (Adams and Boike, 2004; Griffin, 1997). In the latest

Product Development Management Association (PDMA) survey Barczak et al. (2008) finds that stage gate systems are the norm in innovation projects.

The stage-gate model has attracted some criticism. Trott (2008, pp. 409-410) lists five limitations:

1. The process is sequential and can be slow.
2. The process is focussed on the gates and not on the customer so the project teams may find a way of satisfying the internal NPD hurdles without necessarily developing a product or service that is appealing to customers.
3. Product ideas or concepts can be halted too early if the right information is not readily available.
4. The high level of uncertainty that is the hallmark of radical innovation makes this process unsuitable for these types of ideas. Stage-gate processes favour incremental projects.
5. At each stage of the project, a low level of knowledge by the gatekeeper can lead to poor judgments being made on the project.

As noted previously, many companies tailor the stage-gate process to suit the unique or specific intricacies of their business. In their review of the literature, Hauser et al., (2005) suggests that the stage-gate process is often modified by companies practicing design-process management. Companies configure the model to suit their business by having some stages overlap, some activities run in parallel and even dropping certain phases and gates altogether. Breakthrough projects were more likely to be managed using the dynamic, overlapping model whereas low-risk, incremental technology projects used the sequential approach or the conventional stage-gate process regime.

Canner and Mass (2005) also note another shortcoming of the stage-gate philosophy.

While the portfolio model has brought some discipline to the often-undisciplined world of new product development research, the approach has failed to optimise R&D and spark innovation. There may be several reasons for this. Aggressive goals that underpin potential major innovations frequently look "impossible" at project outset, and therefore don't fit neatly into the typical stage-gate review process. The 'organisation invariably shifts attention from the quality of projects to the quantity.' (Canner and Mass, 2005; p. 18)

Stage gate processes, while in widespread use, have been argued, at some level, to be a constraint on creativity (Storey and Salamander, 2005) and can militate against a company pursuing radical ideas. Its overriding focus on process structure, reliability, and control has tended to overlook factors that govern the creativity and flexibility required to innovate (Badaracco, 1991; Moenaert et al., 2000). Stage-gate approaches to innovation have led to a mechanistic interpretation and focus on process efficiency, which is disposed to downplay how process factors such as flexibility, informality, feedback, and autonomy might influence innovation (Clark and Fujimoto, 1991; Dougherty, 1992; Griffin, 1997a).

Goffin and Mitchell (2010) report a cyclical appeal for stage gate processes. Such processes are; 'put in place, often following a bad experience with projects running out of control' (p. 254). In these circumstances, they note, that the introduction of the process is initially welcomed. However, over the course of time, as more is learnt from using the process; more and more gates or hurdles are introduced, resulting in the process itself becoming overly formal and cumbersome. This is compounded by the fact that the organisation is getting better at innovation throughout the process and so managers start to question whether the all the detail and formality is actually worthwhile.

Eventually the process falls into disuse. Then another crisis occurs and it is reinstated, usually in simpler form. (p. 254)

It is therefore not surprising that recent research on the innovation process has acknowledged the need for alternative approaches. Researchers also challenged the assumption that NPD activities could exclusively be represented as an ordered, predictable and sequential system of discrete stages (Kline and Rosenberg, 1986; Leonard-Barton, 1988; Schroeder et al., 1989). Rothwell (1992) argued that the study of NPD as an automatic, dependable, and routine decision-making process did not explain how radical (Dewar and Dutton, 1986; Leifer et al., 2000) innovations emerge.

Reflecting these arguments, researchers have expanded the characterisation of the innovation process as not simply a linear model. Cunha and Gomes (2003) introduce five different product development processes ranging from a linear, sequential model to what they refer to as an improvisation model; one, which has constantly shifting and fluid conditions. They believe that the focus should not be on organisation but on organising. Consistent with this, McCarthy et al (2006) acknowledge that NPD progresses through a series of stages, but with overlaps, feedback loops, and resulting behaviours that resist reductionism and linear analysis. Quinn (1985) was among the first to refer to innovation as a form of controlled chaos.

In summary, stage-gates bring structure to what is, by necessity, a chaotic process. However, the structure alone is insufficient. Strong ideas, that are ideally the fusion of leading technology with compelling customer insights are at the heart of the innovation effort. Any process, to be useful, needs to be able to nurture and bring these ideas to the market. The process must always be secondary to the organisational strategy and the innovation imperative for that context. The stage-gate process should never become an end in itself and this is a real danger because it provides solid framework for organisations and many firms prefer to deal with certainty. As Loewe and Chen (2007) cautioned, while there is enormous benefit in managing innovation like a process with schedules, milestones etc.

> But sometimes, an overly eager project manager intent on making sure that everything gets done on schedule will let the calendar take precedence over the content and the quality of the outputs. (Loewe and Chen, 2007, p. 24).

2.2.5 The Network Perspective

An alternative perspective on innovation examines the role of networks in the process. Studies of the innovation process, partly prompted by the remarkable performance of Japanese companies in world markets, started looking at an interactive, networked perspective for managing innovation (Ahmed and Shepherd, 2010). Takeuchi and Nonaka (1986) investigated NPD within a number of successful companies and reported that the dominant model was not one with linear, sequential phases but one with overlapping ones. Using a sporting metaphor, they communicated the fluidity of the process and the need for different hard and soft skills to facilitate it: 'Stop the relay race. Take up Rugby' (p. 137)

This indicates that a team working on innovation must find its own way through the process. Rather than passing the baton at the end of each discrete section, the new paradigm suggested that players constantly pass the lead back and forth.

This perspective sees innovation as a complex adaptive system (CAS). Complex adaptive systems dynamically try to adapt to the environmental circumstances in which they find themselves. Such systems are found in rich abundance in natural sciences and examples are often drawn from such things as viruses or ant-colonies. Thus, in a way that many organisations would envy, they are able to undertake short term exploitation activities as required and to invest in longer term exploration as needed. The foundation of the CAS framework is based on an understanding that NPD processes are systems whose elements, here called agents, are partially connected and have the capacity for autonomous decision making and social action, known as agency. The decision rules, interactions, and outcomes of agents create three mutually dependent characteristics that define CAS: nonlinearity, self-organisation, and emergence. Carlisle and McMillan (2006) suggest that what innovation should learn from the CAS approach is:

> The message here for organizations is not to take too rigid a stance in approaches to innovation, but to respond flexibly as internal and external environments demand. (p. 4)

In common with all living systems, complex adaptive systems are either connected or in perpetual pursuit of connection. Lewin (1993) underscores the importance of connectedness in evolution and notes that in living systems, ossification occurs without it. Similarly, firms need to ensure that they connect both externally and internally or they may be susceptible to the corporate equivalent of ossification; irrelevance.

Recent studies have demonstrated that innovation mostly emerges through unpredictable, often chaotic processes involving multiple actors and could be better described as a social progress rather than a technical one (Eisenhart in Hargadon, 2003, p. viii). The process of moving from ideation to implementation or launch of an idea is constructed by establishing networks, alliances between human and non-human actors (Christiansen and Varnes, 2007; Kidder, 1981; Kreineer, 2002; Kreiner and Tryggestad, 2002; Callon, 1986; Van de Ven et al., 1999; Takeuchi and Nonaka, 1986).

Hargadon (2003) found that innovation often lies in bringing new knowledge from one area to another by what he refers to as 'technology brokering'. Technology brokering is used to exploit the networked nature of the innovation process. Rather than attempting to pioneer a technological breakthrough in a specific area, technology brokering facilitates the bringing together of people, ideas and combining technologies and objects in a way that sparks new thinking. An example of this type of networked innovation is Thomas Edison, inventor of the incandescent light bulb and the phonograph. Often described as an inventor (Hargadon, 2001) asserts that Edison was, in fact, a technology broker and network builder. He depended heavily on the close collaboration of about 15 people in his R&D lab. Hargadon (2001) asserts that the importance of this collective, creative network can best be judged by the fact that Edison's genius evaporated as soon as the lab ceased to exist. Not only do dedicated R&D labs work this way, but leading consultancy practitioners in the area of innovation also adopt this approach (Hargadon, 2003; p. 200) including the leading innovation company IDEO (www.ideo.com).

New ideas are constantly in search of allies or champions. If nobody adopts a new idea; if nobody advocates it, develops a network to promote and develop it or creates channels for its diffusion, it simply won't succeed. This process includes the involvement of a number of actors in an ever expanding or reducing network. As Tollin and Caru argue: 'Innovation is the relations *in* the network constructed in a dialogue, a constant iteration between human and non-human actors.' (2008, p. 83). The network expands or collapses depending on the input of the various actors connected with the idea and depending on the idea's phase of development.

One key difference between the network perspective and the stage-gate process surrounds the decision making process involved. There are no set, standard points where Go/No-go decisions are mandated. Instead there are a number of heterogeneous, non-sequential, micro decisions that span the network and facilitate the progress of the project. The outcome of the project is contingent on the strength of support it has engaged and the power of the network that has been established during the process. Success is dependent on the number of allies that the project has garnered through the vision it promises and the value the network ascribe to it.

The network process also implies that a team leader will have been able to articulate a compelling vision for the project and that such a vision will attract talented people with relevant skill sets to become involved. One important feature that can drive team or individual motivation is the ambition to do something that has not been done before. However, an idea can travel in many different directions through the network approach and it can be modified, moulded, adapted and altered depending on the relationships that emerge. These relationships are created by a process of 'interessement' where the leader identifies other relevant and useful actors and engages their interest in the problem to be solved/ the innovation project (Akrich et al, 2002).

The idea of extended and networked enterprises (i.e. organisations with many external linkages) has long been the subject of general management research. Networking with other organisations, regardless of the duration of the collaboration, can result in an organisation's attitude to and competence at innovation changing (Hadjimanolis, 2000; Kandampully, 2002; Pavitt, 2002; Flor and Oltra, 2004; Jaskyte and de Riobo, 2004; Medina et al., 2005; Mudrak et al., 2005). This is because the organisation becomes open to new ways of thinking and doing, and learns from the experiences of other organisations or external bodies.

These networks are both theoretically and actually important. It has been noted that the locus of power in innovation projects can transfer from within the original organisation and be found within the network (Powell, 1998). Conway and Steward (2009, p70) note that networks do not have convenient natural boundaries and stretch beyond the confines of an organisation, an industry, a country or even a sector. They distinguish between two types of networks; 'social networks' between individuals and 'organisational networks' between organisations. They provide an example of an organisational network from the electronics industry which shows the diversity and size of an innovation network as well as demonstrating its geographic spread (Figure 2.4).

Figure 2.4: Innovation Networks

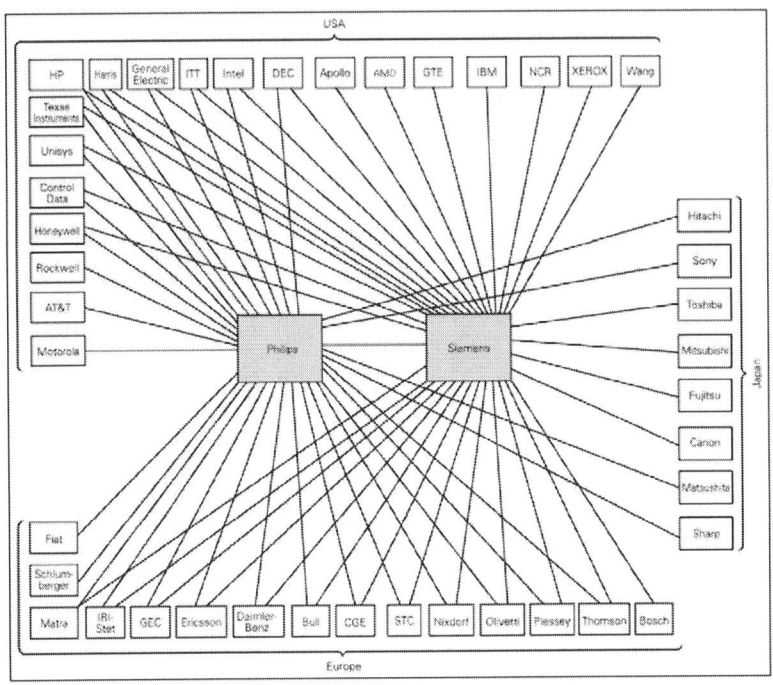

Source: The Focal Network of Philips and Siemens (Contractor and Lorange, 2002)

2.2.6 Open Innovation

The network perspective on innovation reflects an Open Innovation (Chesbrough, 2003) paradigm. In the past, the innovation process was seen as an internal, integrated model carried out by large firms. Innovations were discovered, developed and delivered or launched internally described in the so-called 'Chandlerian' model (Chandler, 1977 and 1990). This model has been recently labeled as "closed innovation model" (Chesbrough, 2003). The 'Chandlerian firm' that rose to prominence in the 20th century has several defining characteristics among which is its self-reliance and expansion into the upstream and downstream activities that more modern firms might tend to outsource. Termed 'giant enterprise', the 'Chandlerian' firm has had significant impact on large sectors of advanced economies and has created an image of what an economic organisation is and ought to be (Robertson and Verona, 2006). By contrast, Open Innovation (OI) considers the boundary between the company and the surrounding ecosystem as porous. Perkins (2008) defines open innovation as the leverage of capabilities and expertise of others to deliver differentiated and meaningful innovation. The open innovation philosophy stems from a recognition that issues such as the speed of advances in technology, combined with global mobility in workforces, in ideas and in venture capital has made it more difficult for conventional R&D teams to be able to contain their knowledge. Moreover, significant innovations are increasingly occurring in small firms, in research labs and institutes or global innovation clusters.

Bessant and Tidd (2011) show that OECD countries are spending over $750bn on research every year through which they are creating a 'sea of knowledge' (p.348) which would be impossible for any single organisation to match – or even stay abreast of. They further point out that such knowledge is no longer merely being created in the advanced industrial nations but that many of the rapid advances are taking place in the developing economies like China and India. They note that without external networks, organisations, however large will inevitably miss out on some key development in their industry.

One of the commercial executives often associated with bringing open innovation into more widespread practice is AG Lafley, then CEO of Proctor and Gamble (P&G) who believed that although P&G had 12,000 scientists and engineers working in R&D when he took over in 2000; he acknowledged that there were many people with great ideas who didn't work for P&G. He set about the purposeful planning of an open innovation culture in P&G which included the rebranding of the existing R&D department to the new name of C&D (for 'connect and develop') thus signalling that P&G was open for business and was actively seeking external partners to develop their ideas (Huston and Sakkab, 2006).

Concurrent with the adoption of innovation networks in business practice, there has been a parallel and distinct body of literature focused on "open innovation" (Chesbrough, 2003). This literature concentrates on inter-firm co-operation and the development of an ecosystem of firms, sharing technologies and trading intellectual property, within a given industry or sector (West et al., 2006). One of the key ways that organisations can increase their external linkages is through their employees having contact with external bodies such as universities and professional institutions (Smith et al., 2008). Open Innovation includes both outside-in (buying) and inside-out (selling) flows of technologies , knowledge and ideas. This model assumes that competitive advantage is inextricably linked to the management of inter organisational relationships with numerous actors external to the firm (customers, universities, competitors, suppliers, research institutions) with the objective of exchanging knowledge, expertise and technological know-how.

2.2.7 The Innovation Value Chain

The two broad, approaches to innovation (linear and networked) describe ways to conceptualise the process; while the innovation value chain outlines the specific phases that are highly likely to occur within every innovation project regardless whether that project has been originated via a development in technology or through a new customer insight, or through a fusion of both. The steps or phases described in the value chain would occur regardless of the overall approach taken.

To improve their firm's innovation performance, managers are advised to view the process of transforming ideas into tangible innovation assets as an integrated flow. For any organisation, to be successful with their innovation activities, its leadership must first, formally define the innovation system and process which is to be followed or used and then apply appropriate quality and innovation metrics and principles; just as was done in the development of quality management, safety management and even finance management in previous decades (Hansen and Birkenshaw, 2007; Barczak et al., 2008). While there is no universal guarantee of success at innovation (Thamain, 2008), companies can stack the odds in their favour by having structures and processes and by using some formal process to manage their innovation activities (Schmidt et al., 2009). Based on this formulation, O'Reilly and Flatt (1989) concluded that innovation results from two component processes:

(a)   creativity or the generation of a new idea, and

(b)   implementation or the actual introduction of the change.

This suggests that the enhancement of innovation in organisations requires mechanisms for both stimulating new ideas as well as methods for putting the ideas into practice. More recent literature has evolved this two factor framework into three elements. The addition is a stage within the innovation process within which ideas are sifted, ranked, prioritised and where raw ideas are incubated and transformed into testable concepts.

O'Connor and Ayers (2005) advocated a three part programme for innovation in which the three elements are discovery, incubation and acceleration. Such a three-part division of the innovation process is increasingly a feature of this literature (e.g., Cooper and Kleinschmidt, 1993; Veryzer, 1998; Tidd and Bodley, 2002; O'Connor, 2005; Vuola and Hameri, 2006; Hansen and

Birkenshaw, 2007; Roper et al., 2008). The three parts described are generally configured thus: a) the discovery or idea generation phase; b) the incubation or transformation phase and; c) the commercialisation or implementation phase.

Regardless of the source of the original idea, this model assumes that all ideas are born as relatively raw, fragmentary, embryonic thoughts and connections and they require some level of incubation in order to develop the nascent idea into a testable concept. They are essential blocks in accomplishing the goal of new product or service development. In contrast to the stage gate model, the innovation value chain does not assume that there are decision gates at intervals during the process that dictate whether individual projects will receive further funding and be continued. The innovation value chain describes steps in a chain of activities that transcend the organising model or approach used.

In the first stage; idea generation, new ideas and opportunities are identified and new technologies evaluated. In the incubation or transformation phase, selected ideas are converted into testable concepts and evolved into business plans with proof of concept established and prototypes developed. In the commercialisation, implementation or diffusion phase, the prototype moves out of the R&D department and the overall organisation starts to put additional resources and demands on the new business idea.

Hansen and Birkinshaw (2007) recommend viewing innovation as a value chain comprising these three key phases: idea generation, (idea) conversion and (idea) diffusion. Figure 2.5 illustrates the Hansen and Birkinshaw (2007) Innovation Value Chain. The innovation value chain is derived from the findings of five large research projects on innovation that Hansen and Birkinshaw undertook over the past decade. They interviewed more than 130 executives from over 30 multinationals in North America and Europe. They also surveyed

4,000 employees in 15 multinationals, and they analyzed innovation effectiveness in 120 new-product development projects and 100 corporate venturing units. This firm-level perspective is more deterministic, connecting organisational performance and innovation success to company knowledge and internal resources. This view is consistent with a resource-based or capabilities perspective on innovation and new product development (Foss, 2004).

Figure 2.5: The Innovation Value Chain

	IDEA GENERATION			CONVERSION		DIFFUSION
	IN-HOUSE	CROSS-POLLINATION	EXTERNAL	SELECTION	DEVELOPMENT	SPREAD
	Creation within a unit	Collaboration across units	Collaboration with parties outside the firm	Screening and initial funding	Movement from idea to first result	Dissemination across the organization
KEY QUESTIONS	Do people in our unit create good ideas on their own?	Do we create good ideas by working across the company?	Do we source enough good ideas from outside the firm?	Are we good at screening and funding new ideas?	Are we good at turning ideas into viable products, businesses, and best practices?	Are we good at diffusing developed ideas across the company?
KEY PERFORMANCE INDICATORS	Number of high-quality ideas generated within a unit.	Number of high-quality ideas generated across units.	Number of high-quality ideas generated from outside the firm.	Percentage of all ideas generated that end up being selected and funded.	Percentage of funded ideas that lead to revenues; number of months to first sale.	Percentage of penetration in desired markets, channels, customer groups; number of months to full diffusion.

Source: Hansen and Birkinshaw (2007)

Hansen and Birkinshaw (2007) suggest that executives need to view the process of transforming ideas into commercial outputs as an integrated flow, from end-to-end. The first of the three phases in the chain, they describe, is to generate ideas; this can happen in three ways. Companies can develop ideas within a single department, or across the company using cross-functional teams or they can involve external partners to generate the ideas.

The first phase is linked to organisational creativity. Any new product development (NPD) process requires a high level of creative performance. According to Leenders et al (2007), creative performance is of paramount importance in NPD projects and most NPD projects are managed through an NPD team as the organisational nucleus for innovation. Innovation inevitably involves creativity: the initiation, identification or discovery something novel, an idea, technology, or process that is new to the organisational setting which is then followed by its development and implementation (e.g. Amabile, 1988; Dougherty and Hardy, 1996; Kanter, 1988; Klein and Sorra, 1996).

Despite its importance, creative performance of innovation teams, however, is scarcely examined and knowledge about the conditions that enhance or obstruct innovation teams' creative performance is limited (Kratzer, 2004). One factor that has been recognised to impact the creative performance is the team's communication (e.g. Leenders et al., 2003) and team communication also involves, to a certain extent, disagreements between communicating colleagues.

The second phase is to convert ideas; to incubate the best ones and to amplify the elements of the ideas that have most appeal. More specifically, the second phase helps select, sift, rank and prioritise ideas for funding (or resourcing) aimed at developing them into products, services or practices. The third phase is to diffuse, exploit or implement those products and practices both inside the organisation or outside in the case of launching new products and services or creating new markets.

When organisations are asked to rate their innovation capability using the value chain model, Hansen and Birkinshaw (2007) assert that firms typically fall into one of three broad "weakest link" scenarios. First is the idea-poor company, which spends a lot of time and money developing and diffusing mediocre ideas that result in mediocre products and moderate financial returns. The problem is in idea generation, not execution. By contrast, the conversion-poor company has lots of good ideas, but their leaders don't screen and develop them properly. Instead, ideas can perish in budgeting processes that 'emphasise the incremental and the certain, not the novel.' Other managers subscribe to the "1,000 flowers" approach, letting ideas bloom where they may but never weeding the weaker ones out. The need here is for better screening capabilities, not better idea generation mechanisms. Finally, the diffusion-poor company has trouble launching, placing its bets and profiting from its good ideas. Decisions about what to bring to market are made locally, and not-invented-here thinking can prevail. As a result, new products and services may not be comprehensively rolled out across geographic locations, distribution channels, or customer groups.

Roper et al (2008), developed a similar model in which an innovation event, like the launch of a new product, service or process, represents the end of a series of knowledge sourcing and translation activities by a firm. It also marks the start of a means of value creation that, subject to the firm's capabilities and the buoyancy of the markets it operates in, should yield an improvement in NPD results. According to Roper et al. (2008), the first link in the innovation value chain is a firm's knowledge sourcing activity; these authors focus in particular on the factors that drive firms' engagement with particular knowledge sources; experts, research institutes etc. The second link in the innovation value chain is the process of knowledge transformation, in which knowledge sourced by the enterprise is translated into innovation outputs. This is modeled using an innovation or knowledge production function in which the effectiveness of a firm's knowledge transformation activities is influenced by enterprise characteristics, the strength of the firm's resource-base, as well as the firm's managerial and organisational capabilities.

The final link in the innovation value chain is knowledge exploitation, i.e. the firms' ability to fully commercialise their innovations. While this model builds closely upon the Hansen and Birkinshaw (2007) innovation value chain model, it does contain some specifics about how and, specifically, where firms can access knowledge that may be useful as a start point for new product or service ideas. The authors classify five sources of such knowledge: Internal dedicated

R&D; backward linkages to suppliers and consultants; forward linkages to customers/consumers; horizontal linkages to competitors or joint ventures and public linkages to research institutes and universities.

The Roper et al. (2008) model maps well against the Hansen and Birkinshaw (2007) Innovation Value Chain (IVC) model (Table 2.1). Knowledge sourcing is equivalent to idea generation: transformation is comparable to conversion and exploitation and diffusion are also similar.

Table 2.1: Phases in the Innovation Value Chain

Phases	Stage 1	Stage 2	Stage 3
O'Connor and Ayers (2005)	Discovery	Incubation	Acceleration
Loewe ad Chen, (2007)	Discovery	Opportunity	Realisation
Hansen and Birkenshaw (2007)	Idea Generation	Idea Conversion	Idea Diffusion
Roper et al (2008)	Knowledge Sourcing	Transformation	Exploitation

It is worth noting that dividing the innovation process into an innovation value chain is not the same as creating a stage-gate process. The stage-gate is far more prescriptive and mechanistic. It mandates a sequence of defined activities punctuated by key decision points. The value chain is an indicative model to guide firms and teams in the key activity of facilitating innovation by knowing 'where they are in the game.' (Amabile and Khaire, 2008, p. 104). The gate process is the dominant linear model of innovation and it maps the flow of decisions that may punctuate the progress of an innovation project. The Value Chain, on the other hand, explores the individual activities that must be discharged within any specific project.

2.3   The Nature of Innovation – What Makes it Such a Challenge for Managers?

2.3.1   Types of Innovation; Radical and Incremental

Innovations are often analysed in terms of contradictions: *incremental* and *radical* (e.g., Green et al., 1995; McDermott and O'Connor, 2001); *continuous* and *discontinuous* (Veryzer, 1998); and *sustainable innovation* and *disruptive innovation* (Christensen, 1997). Diverse descriptions abound when it comes to defining radical innovation. Some use the label of 'discontinuous innovation' (Anderson and Tushman, 1990), 'emerging technology' (Day and Schoemaker, 2000), 'architectural innovation' (Abernathy and Clark, 1984), or even 'disruptive' technology (Christensen, 2000). For the classification of these types of innovations, a four-box matrix is often used. Such a matrix usually comprises a market or demand dimension (existing and new) and a technology dimension (existing and new). It is widely acknowledged that in terms of new products and services and their impact on both the industry and firm level, significant differences exist between radical and incremental innovations (Dosi, 1982; Christensen and Rosenbloom, 1995; Christensen, 2000).

Figure 2.6: Types of Innovations

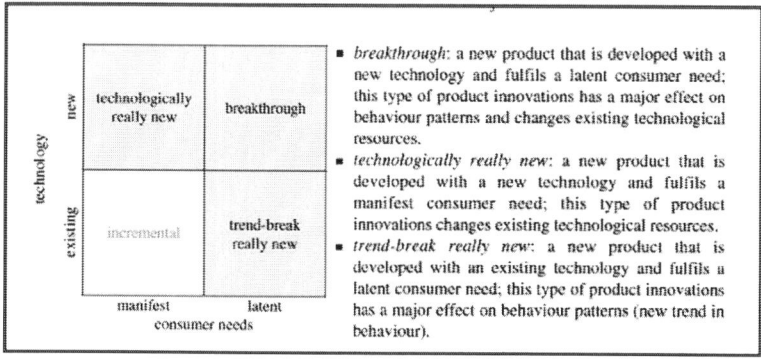

Source: Janssen (2008).

Incremental innovation leads to small improvements in products, services or business processes. Radical or breakthrough innovation, on the other hand, is 'game-changing' and usually results in redrawing the rules of the industry or category. In the above matrix, the consumer need dimension reflects the demand side and is divided into "manifest" and "latent. Manifest needs are needs that can be articulated and described by people; latent needs cannot, at least, not consciously (Narver et al., 2004).

Bessant and Tidd (2007, p. 15) prefer a six-box model to describe the degree of novelty that separates incremental from radical innovation. They use the analogy of comparing innovation to Russian dolls; i.e., you can change things at the (relatively low) level of components or else you can change the whole system. For example, they say, you can put a faster microchip on the circuit board for the graphics display in a computer (component innovation) or you can change the way computers work by creating a network linking pc's to drive a small business (architecture innovation). They note that there is scope for innovation at every level but that changes at the higher levels inevitably have implications for the lower levels.

Figure 2.7: Bessant and Tidd: Incremental and Radical Innovation

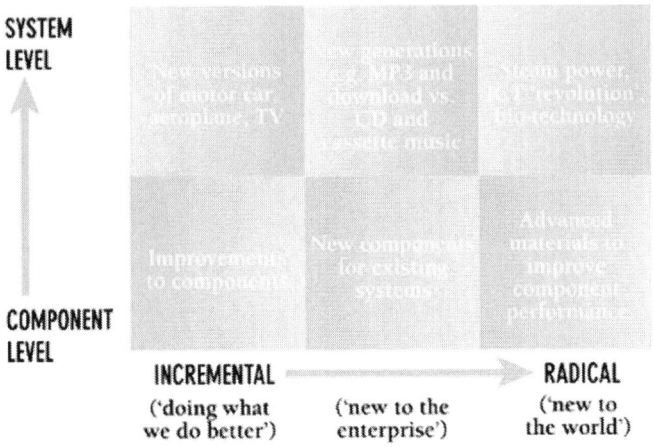

Source: Bessant and Tidd (2007, p. 15)

Garcia and Calantone (2002) noted that the proliferation of different definitions, especially around the classification of 'radical innovation' was giving rise to an unhelpful ambiguity within the literature around new product development:

> A plethora of definitions for innovation types has resulted in an ambiguity in the way the terms 'innovation' and 'innovativeness' are operationalised and utilised in the new product development literature... The terms radical, really-new, incremental and discontinuous are used ubiquitously to identify innovations. One must question, what is the difference between these different classifications? To date, consistent definitions for these innovation types have not emerged from the new product research community. (Garcia and Calantone, p. 110)

Regardless of the variation in words used to describe radical innovations, some common elements are present in most definitions. The definitions generally allude to aspects related to high market and technological uncertainty, new market creation. They often refer also to creating new capabilities in the innovating firm as well as the possibility of cannibalizing the firm's prior business model. Leifer et al. (2001) define radical innovation thus:

> A radical innovation is a product, process, or service with either unprecedented performance features or familiar features that offer significant improvements in performance or cost that transform existing markets or create new ones. (p. 103)

Roberts (1999) suggests that a firm that consistently innovates will demonstrate sustained profitability but that breakthrough (radical) innovations are especially important to organisations. They enable firms to challenge the existing technological order and create new opportunities, new directions. Further, they allow firms to renew, reinvent themselves; to develop new opportunities for business growth (Burgelman, 1983). They represent valuable and worthwhile sources of competitive advantage for firms (Barney, 1991). Therefore, Ahuja and Lampert (2001) argue that understanding the determinants of these type of innovations is critical to the literature on strategy and organisational learning. Gary Hamel (2000) also argues in favour of radical rather than incremental innovation saying;

> Most companies long ago reached the point of diminishing returns in their incremental improvement programs. Radical, non-linear innovation is the only way to escape the ruthless hyper-competition that has been hammering down margins in industry after industry. (p. 36)

In competitive markets, all innovation, but radical innovation in particular, has been argued to be a strategic imperative. Some commentators stress the need for radical 'frame breaking' discontinuous innovation at the expense of incrementalism. For example, Peters (1999) believes that "incrementalism is innovation's worst enemy" (Peters, 1999: p. 27). He suggested that "the only sustainable competitive advantage comes from out-innovating the competition" (p. 29).

Song and Thieme (2009) reviewed radical innovation from a transaction cost perspective and observed:

> Radical innovations rely on both physical and human capital that is less standardized and more transaction specific (idiosyncratic) than incremental innovations. Often, new facilities and equipment are needed to produce a radical innovation. Likewise, market intelligence gathering activities in radical innovation are likely to involve new investments in both physical and human capital that is transaction specific. Firms (or their suppliers) often must develop new relationships with potential customers in unfamiliar market segments or use new methods of gathering intelligence (i.e., lead user analysis) when pursuing radical innovation. (p. 44)

The degree of radicalness of an innovation, however, is conceptually challenging to define or measure. Such categorisation of innovations as either radical or incremental is, for a number of reasons, subjective. It is not one of hard and fast, bounded categories. Instead, there is a continuum of innovations that range from radical to incremental (Hage, 1980). A new product or service's position on this continuum depends upon perceptions of those familiar with the degree of departure of the innovation from the state of knowledge prior to its introduction.

Radical or breakthrough innovations are defined as new products that are the first to bring novel and significant consumer benefits to the market (e.g. Chandy and Tellis, 1998; Sorescu et al., 2003; Zhou et al., 2005). These benefits can range from improvements to product features, such as packaging (e.g. a new type of spout on a paint canister), to opening up an entirely new market (e.g. gambling or spread-betting on the performance of stocks and shares rather than on race-horses). On the other hand, incremental innovations are new products or services that do not deliver novel and significant consumer benefits to the market (like a new flavour or pack size of an existing product or brand). However as Conway and Steward (2009, p. 17) assert, assessing the novelty or extent to which a new product is radical is not a straightforward matter. They offer five perspectives for assessing the novelty or radicalness of an innovation. These categories are not mutually exclusive and overlap to some degree.

The embedded characteristics of the innovation: the degree to which the innovation offers improved technical functionality and performance as well as the quality of the underlying technology that enables this performance level.

Benefit to the user/consumer through usage or consumption of the innovation: the degree to which the embedded characteristics of the innovation provide the user with new benefits or possibilities or changes to their patterns or routines.

The breadth of diffusion of the innovation: the more widely used an innovation is, the greater the collective impact it may have on society and business. Thus, the more pervasive (widely diffused and used) an innovation is, the more likely it is to be perceived as radical. Innovations, like electricity, the telephone, the internet have all been extensively diffused across geographies and business applications.

The impact on an innovating organisation's competencies and capabilities: Radical innovation is innovation that requires new technological knowledge to exploit it, whilst rendering existing technical knowledge, in that area, obsolete. In contrast, incremental innovation uses and builds upon existing technology. Thus radical innovation can be seen as competence destroying.

The time elapsed since the launch of the innovation – The perceived novelty of an innovation decreases over time as it becomes more widely diffused and embedded in patterns of work or, indeed, becomes superseded by some newer innovation.

The issue of novelty is raised in this discussion as central to the notion of radicalness; and novelty is a subtle element of the innovation process. Novelty is generally defined as not necessarily a property of the innovation itself but specifically, novelty is a feature of the perception of the individual or organisation that adopts the innovation. Zaltman et al. (1973, p. 10) see perceived novelty as a central feature of innovation:

> Any idea, practice or material artefact perceived to be new by the relevant unit of adoption.

In the same vein, Rogers (1995, p. 11) suggests that:

> It matters little whether or not an idea is objectively new as measured by the lapse of time since its first use or discovery…If the idea seems new to the individual, it is an innovation.

Simply, this means that the perception of novelty of an innovation by an individual is generally linked to the point in time of adoption, rather than the point in time of the discovery or original commercialisation of that innovation. Ex post, the definitions of radical versus incremental innovations can be based on the effects of an innovation on the technology within that domain, the product assortment, the marketplace, or on the individual firm. Govindarajan and Kopalle (2006), for example, differentiate between the radicalness and disruptiveness of an innovation by arguing that 'disruptive' refers to market disruption whereas 'radicalness' is generally applied to the technology involved. Garcia and Calantone (2002) take a broader perspective and state that radical innovations cause both marketing and technological discontinuities on a macro and micro level.

Radical innovation is increasingly the focus of academic research into the factors that cause or even promote it. A number of contributions have already been made towards developing a model or inventory of correlates to radical innovation. Many of the factors identified are at the corporate level and, as such, difficult for individual managers to influence. Nevertheless, there is a growing body of work making a theoretical contribution to the construct of radical innovation.

McDermott and O'Connor (2002) observe that because of the specific nature of radical innovation, the gestation period for such projects is long. This is because often both the technology and the marketing opportunity are distant from the company's current commercial footprint. Golder et al., (2009) found that the gestation or incubation period for a range of (29) radical innovations they studied to be on average 23 years. This project-longevity means that such initiatives will typically outlast the project team members and the management supporters who decide to commence them. It is highly unlikely that the same management will survive to protect and champion the project and this factor adds to the uncertainty surrounding radical projects. With new management, often comes new priorities and this combined with external market developments add to the on/off, stop/start nature that characterises many of these projects.

Prior research has identified an inventory of factors which are believed to support the outcome of radical innovation. Such factors include a learning and continuous improvement culture (Bessant and Francis, 1997: McLaughlin et al., 2008); lead user involvement (Von Hippel, 1988); project management skills for managing individual projects and corporate systems for managing portfolio's of innovation projects (Cooper, 1984; Wheelwright and Clark, 1992); cross functional teams with high ability to solve problems (Jassawalla and Sashittal, 1999; Sapsed et al., 2002); use of design thinking methods including user-centred design and prototyping (Dodgson et al., 2005; Brown, 2008; Verganti et al., 2007); centralising the team and creating an innovation hub (Ettlie et al., 1984; Leifer et al., 2007); organisational tolerance for ambiguity, failure and risk (Kalunzy et al., 1972; Nembhart, 2009); high levels of both individual and organisational creativity (Amabile, 1998; Tagger, 2002); clear and active senior management support (McDermott and O'Connor, 2002); capacity for deep market insight and an ability to convert weak market signals into actionable innovation projects (O'Connor, 1998); a commercial willingness to engage with projects which have a lengthy gestation period and an associated high cost (McDermott and O'Connor, 2002; Golder et al., 2009) ; team climate for innovation (Anderson and West, 1998; Caldwell and O'Reilly, 2003); firm size and position in the market (Christensen, 2007; Chandy and Tellis, 2000) and personal and corporate networks, alliances and relationships (Bessant et al. 2003; Reed and Walsh 2002; Chesbrough, 2003; Birkinshaw et al., 2007). In Chapter Eight, these factors will be revisited in the light of the case study experiences.

In their book, Radical Innovation (2000; p. 19), Leifer et al, propose a useful and comprehensive inventory of ways in which radical innovation is markedly different from incremental innovation (Table 2.1). The differences generally revolve around the longer timeline associated with radical innovation projects as well as the higher uncertainty attached to them in terms of technological and commercial success.

Table 2.2: Incremental Versus Radical Innovation

	Incremental	Radical

Project time line	Short term – six months to two years	Long term – usually ten years or more
Trajectory	Linear continuous path from idea to commercialisation	Multiple discontinuities, gaps, hibernations and revivals, unanticipated events, outcomes and discoveries
Idea Generation and Opportunity Recognition	Occur at the front end of the process; critical events are largely planned for	Occur sporadically throughout the life cycle of the project often in response to discontinuities of funding, technical, personnel or market
Business Case	Complete and detailed plan is made because of the low level of uncertainty	Business model evolves through the technology and market learning that evolves through the project.
Key players	Assigned to cross functional team, each member has a clearly defined role	Key players come and go and many remain interested through an informal network that often grows around such projects
Organisational structure	Cross functional team within a business unit	Often starts in R&D but may migrate into incubation unit
Operating Unit involvement	Operating units involved from the beginning	Informal involvement with operating units but project actively avoids becoming captive to an operating unit too early

Adapted from: Leifer et al., 2000 (p. 19-20)

2.3.2 Innovation and Risk

It has been suggested that radical or disruptive innovation is especially risky for firms. Indeed, Kalunzy et al. (1972) define radical innovations exclusively in terms of risk. This risk is associated with a lower likelihood or speed of product adoption or a lower likelihood of survival of the innovating firm. Researchers have even suggested that these products often fail in the marketplace because managers consistently overvalue their benefits relative to existing products, whereas consumers systematically undervalue them or forgo them altogether in favour of more familiar products (Gatignon and Robertson, 1985; Gourville, 2005).

There is a general acceptance of innovation's importance, by the leadership of most companies, but yet there is also an accompanying general dissatisfaction with the results accomplished by innovation investments (Dervitsiotis, 2006). Crucially, there is a difference between having innovation as a priority and having it as a capability (Hamel, 2005). Sorescu and Spanjol note that many executives hold an unassailable belief in innovation as *the* ultimate strategic imperative, relying on it to spur growth and yield positive financial returns

(2008). However, they say, profitable innovation remains an elusive goal. Some researchers posit that simply innovating more will generate superior performance (Bayus et al., 2003; Pauwels et al., 2004). Others argue that incremental innovations, which make up 90% of new product introductions, have little or no impact on firm value (Christensen, 1997; Foster and Kaplan, 2001). Cooper called new product development (NPD) "one of the riskiest, yet most important endeavours of the modern corporation" (1993, p. 4). Davila et al. concur that more innovation is not necessarily better (2008, p. 17). They contend that 'innovation, like most things, is best in the right proportions.' (p. 18)

Radical innovation, as noted previously, has traditionally been defined as that innovation territory where both technical and market uncertainties are high. Technical uncertainties refer to questions about the appropriateness and application of the underlying scientific knowledge, whether the technology is fit for purpose, whether the technical specifications of the product are right and whether it can be scaled up and made. Market uncertainties include issues related to customer needs and wants; to what extent consumers will adopt the new product or service and how the distribution channels will block or enable the diffusion of the idea. To these two uncertainties, Leifer et al. (2002) add a further two; resource uncertainties and organisational uncertainties.

> Among the organisational uncertainties were questions about the capabilities of the project team; recruiting the right people; managing relationships with the rest of the organisation; dealing with variability in management support; overcoming the short-term, results-oriented orientation of operating units, and their resistance to products that might jeopardise existing product lines; and counteracting vested interests in the current business model. (p. 104)

Resource uncertainties refer to the availability of physical, financial and human resources to get the necessary things done. These were found in the Leifer et al (2002) study to require an inordinate amount of time from the project team. In fact, they concluded that the greatest contribution that organisations could make to the development of radical innovation is to manage the internal organisational uncertainties that are within their control.

> If firms learn to reduce these uncertainties in a systematic way—through leadership and organizational and managerial approaches—then radical-innovation project teams would be better able to address the less controllable and more chaotic market and technical uncertainties. (Leifer et al., 2001; p. 104)

Furthermore, in markets started by a radical product, the first firms to enter the market have the lowest survival probability (Min et al., 2006). Perhaps consistent with this view, the numbers of breakthrough products or services launched in recent years have been declining (Jannsen, 2008).

More attention is being paid recently to discontinuous, disruptive innovation because empirical studies have continued to identify a worrying pattern of results for large organisations. Tushman and O'Reilly report this trend in almost every industry studied including automotive, cameras, colour televisions, optical equipment, hand tools, stereo equipment, tyres and watches (2007). The incumbent market leader fails to maintain industry leadership during a sustained period of discontinuous change (Paapa and Katz, 2004, pp. 1-3). Dosi (1981) stated that it is new, small firms that rapidly emerge as key players in a sector when there is a "paradigm shift" in technology, which alters radically the rate, direction and skills associated with a technological trajectory. Similarly, Bessant et al., noted that most radical innovation in mature industries was emanating from start-up businesses (2004). Organisations who have a dominant share of lucrative markets must ensure that they are not toppled by more agile, often smaller competitors who develop radical innovations in those markets (Christensen, 1997; Bower and Christensen, 1997).

2.3.3 Some Common Barriers to Innovation Inside Organisations

The fact that innovation demands creative, non-routine responses makes it difficult to design programmed actions that are more likely to deliver innovative outcomes a priori (Caldwell and O'Reilly, 2003). Leenders et al. (2007) concur that companies are seeking to balance the largely loose and unsystematic conditions, under which creativity thrives, and the simultaneous need for more systematic design methods to reduce the risk and the costs of the project. The concern is often voiced that superimposing a system or predefined process onto the NPD process can stifle or hamper the necessary creativity.

The idea of an innovation journey was proposed by Van de Ven *et al.* (1989) to illustrate the complexities, barriers, contingencies and uncertainties of new product, process and service development processes. Innovation processes are seen as expeditions into unexplored territories, which are "highly ambiguous and often uncontrollable and unique to its travellers" (Van de Ven et al., 1999, p. 21). Emphasising the ambiguity of innovation initiatives, Markides (2002) outlined an inventory of elements that conspire against established firms in pursuing innovation:

> Compared to new entrants or niche players, established companies find it hard to innovate because of structural and cultural inertia, internal politics, complacency, fear of cannibalising existing products, fear of destroying existing competencies, satisfaction with the status quo, and a general lack of incentive to abandon a certain present (which is profitable) for an uncertain future. (p. 246)

March and Simon (1958) discussed the opposing demands of exploration and exploitation in management and leadership of firms. Exploitative activities are usually dominated by mechanistic structures, predictability and routine, all of which help drive efficiency. Processes are rigid and are optimised for activities that lead to a better performance in a known environment (Tushman & O'Reilly III, 1996). Explorative activities, on the other hand, are characterised by more openness; they are more exploratory and entrepreneurial and consequently come with higher risk for failure – but also with the potential for higher success. Exploring new opportunities while simultaneously exploiting current opportunities remains a significant management challenge.

Doughtery (1992) speculates that for some organisations, the challenge will be too great:

> Despite the importance of product innovation, research has shown that established organisations have difficulty in developing and marketing commercially viable new products. (p. 77)

> Can large, old firms, in fact, change their fundamental principles of management or must they "die" to make way for new forms? (p. 90)

Large organisations are arguably more successful at developing incremental innovations in familiar product groups and have been notably unsuccessful in being able to commercialise breakthrough ideas. O'Reilly and Tushman note this phenomenon to be both commonplace and also fascinating (2004). They built on the description of an ambidextrous organisation as one which can simultaneously focus on the operational requirements of today's business while also seeking and building the transformational opportunities for the future. Academics have acknowledged the requirement for firms, especially those in a highly competitive environment to combine both explorative and exploitative activities in a single business unit (Benner & Tushman, 2003; Gibson & Birkinshaw, 2004; Jansen et al., 2005). This type of management has been described as contextual ambidexterity. Jansen et al., argue that firms who follow it have to combine opposing coordination mechanisms that include decentralisation, formalisation and connectedness (2005). It has been found that this type of working requires high intellectual capabilities of the individuals within the innovating units. Individual managers within a firm or team have to decide themselves when and how to perform exploitative or explorative activities (Gibson & Birkinshaw, 2004). Hence, the individual managers have a high influence on the innovativeness of their group (Subramanian & Youndt, 2005).

In this context, research has shown that it is often the managers themselves who are among the most significant barriers to innovation within firms and within projects. Some see innovation as risky and subversive and, instead of supporting it, they see their role as needing to protect the organisation from innovation (Storey and Salaman, 2005). Culture (Pech, 2002) plays a role in the behaviours of managers which could be described as anti-innovation:

> A dominant culture of conformity and followership generates "more of the same", while a culture encouraging individualism and leadership produces new products or methods of production by harnessing employee creativity and innovation. (p. 559)

Storey and Salaman (2005) were able to classify a number of companies into high and low performing, in terms of innovation. For those whose performance was shown to be below average they found an attitude in which innovation was characterised as: inherently messy; hard to control; childish, creative; threatening and unsettling – even dangerous and irresponsible but not essential. Contrastingly, in the above average performing companies they found innovation was seen as quirky; It has the capacity to break down barriers and bring new, useful ideas to the surface; it's exciting; it's exhilarating and it's a key source of competitive advantage and central to success.

2.4   Innovation in Teams

According to Kratzer et al. (2004), teams constitute the organising principle in most modern, innovation companies. A key source of the innovation initiatives of companies is the creative performance of the people they employ (Cummings and Oldham, 1997). One particular organisational response, the implementation of cross-functional teams, has progressively become a more common approach in the corporate quest for enhanced levels of innovation (Cox et al, 2003; Pearce and Sims, 2002). Empirical research has shown that in 1997, 80% of (US) firms with more than 100 employees used a team based approach to innovation (Cohen and Bailey, 1997).

A number of reasons have been suggested as to why teams are the organising nucleus for innovation. Innovation is often said to take place on the borders or fringes between disciplines, or specialities, when people start to share ideas with each other. These interactions are vital because they enable individuals to combine different understanding and experiences to generate and bring to the surface perspectives that may be different and more valuable than those held by any individual team member. New product and service development is dependent on organisation members' ability to combine and exchange knowledge (Smith et al., 2005) and hence teams are considered a natural and powerful option for creating and circulating innovative ideas.

Edmondson (2002) argued that innovation inherently occurs at the team level because it requires learning behaviours, or transmission of knowledge that takes place through conversations between a limited number of interdependent people. These interactions are vital because they enable individuals to combine different insights and experiences to generate and bring to the fore knowledge that is greater than that held by any single individual member of the team (Nonaka and Takeuchi, 1995), in the sense that the whole is greater than the sum of the parts.

Working on innovation and NPD necessitates highly integrated teamwork (Dym and Little, 2000). Teams, like individuals, process information by encoding, storing and retrieving it (Brauner and Scholl, 2000). But, unlike individuals, by communicating, teams build on the knowledge of others, exchange information experiences and insights and build new knowledge and create new ideas (Leenders et al., 2007). Frequency of team communication (Kratzer et al., 2004) is also a mediating factor in the success of NPD projects. There is a heightened level of communication within teams where no formal approach is necessary between members (such informality is unlikely to be the case if the individuals are not connected by the team) and casual conversations, information exchange and discussions may emerge more naturally and easily.

In essence, teams can offer more flexibility, productivity, and creativity than any one individual can offer (e.g. Gladstein, 1984; Hackman, 1987) and provide more complex, innovative, and comprehensive solutions to organisational problems (Sundstrom, DeMeuse and Futrell, 1990). New product and service development is dependent on organisation members' ability to combine and exchange knowledge (Smith et al., 2005) and hence teams are considered a natural and powerful option for creating and circulating innovative ideas (Jackson, 1992; Denison et al., 1996; Donellon, 1996; Griffin, 1997; Thompson, 2003).

Although many people see creativity as essentially an individual practice (Amabile, 1998), empirical research findings by Taggar (2002) suggested that team level innovation processes are actually needed to bring individual creativity into use and that, without team level interactions, insights and efforts, these processes may be carried out in vain with no firm-level benefits emerging. Similarly, although individual employees can develop innovations, teams of employees will be more important in influencing overall ability of the organisation to innovate (Anderson and West, 1998; Read, 2000; Lemon and Sahota, 2004; Noke and Radnor, 2004).

Muthusamy et al. (2005) argue that as organisations are increasingly emphasising innovation, self-managing work teams (SMWTs) can be expected to facilitate the transition towards an innovative workplace. SMWTs are often associated with high performance work systems, and their use is more widespread than any other flexible work practice (Osterman, 1994). Such teams are responsible for carrying out their work as well as for supervising their team's own performance. Instead of having a formal line manager instruct them, these teams are responsible for gathering information, making decisions, and meeting the organisational goals (Hollander & Offermann, 1990).

Innovation is facilitated by organic structures and flexible work arrangements characterised by autonomy, higher degree of informality, intense information exchange, and participative decision-making (Taggar, 2002; Collins & Amabile, 1999; Kanter, 1988). Since SMWTs demonstrate many of these characteristics, they can be expected to have major a impact on innovative behaviour (Dunphy & Bryant, 1996).

2.4.1 Innovation within the R&D Context

R&D is often the core activity and starting point for innovation (Kratzer, 2006). The R&D setting for innovation deserves specific consideration and treatment for three reasons: first, the setting is unique; second, the type and nature of the work is specific to that setting; and, third, the people engaged in R&D are also uncommon in many respects. R&D settings are unique insofar as quality rather than quantity and innovation rather than cost are the primary performance criteria (Keller, 1989).

A concern voiced in the R&D management literature is the short-term focus of senior management. Mitchell and Hamilton (2007) noted a widespread sense that corporate strategy and short term budget discussions frequently discriminated against longer-range and higher-risk R&D initiatives. By insisting that investments in R&D meet the same financial criteria that apply to all other investment decisions within the corporation Mitchell and Hamilton (2007) suggest that the R&D community felt that this over emphasis and misplaced faith in financial ratios results in the rejection of many of the programs and lines of inquiry that held out the most promising opportunities. This phenomenon make R&D a more difficult context in which to work as many of the most attractive projects are likely to be the very ones in high danger of abandonment because they will be the very ones which it is likely to be most difficult to justify on a return on investment basis.

In terms of the type of work involved in R&D, Mumford et al. (2002) characterise the creative work of R&D as highly demanding, time-consuming, resource intensive and requiring a high level of persuasion and politics. The tasks developed in this kind of work are uncertain, very risky and involve unforeseen processes, and as a result, setbacks and disruptions are likely to occur (Kim et al., 1999). The very nature of R&D suggests that its outcomes are likely to stretch the boundaries of what is known in a certain domain. Therefore, the process will, by definition, lead teams into areas of novel technology where uncertainty is very high and where the safety of routines, best practices and established ways of working are, necessarily, absent because they have not yet evolved.

Many teams working in the industrial research or pharmaceutical R&D setting exist specifically to produce major, novel, and creative innovations that might be considered new to the world in scientific, technical or medical areas. A distinction is traditionally made (e.g. by Bain et al., 2001) between *research*, which encompasses fundamental, basic, exploratory and applied research requiring high levels of new knowledge creation, and *development*, which covers technical service, product, and process development work involving the application of existing knowledge to solve particular problems (Leifer and Triscari, 1987).

In sequence, research occurs before development. It is the output of the research phase that is normally handed over to the development team to 'make it happen' or 'bring it to life'. This conforms to the traditional linear view of innovation (Cooper and Kleinschmidt, 1987), where ideas emerge in an environment of relative creative freedom and get gradually refined over time to the degree that they become more tangible and capable of development ultimately into products and services or tangible innovation assets.

This process is often characterised as being inherently uncertain, dynamic, and to follow a seemingly random process (Kanter, 1988; Jelinek and Schoonhoven, 1990). The recognition and acknowledgement of the element of chaos was chronicled twenty-five years ago by Quinn (1985). Cheng and Van de Ven (1996) found that the actions and outcomes experienced by innovation teams tend to exhibit a chaotic pattern during the initial period of innovation development (R), the research phase, and an orderly, regular pattern during the concluding development (D) period. This seems inherently intuitive as the 'R' phase is the most creative and the 'D' phase requires that existing ideas (coming out of the 'R' phase) are developed into tangible innovation assets which are capable of undergoing research and proof of principle tests.

Apart from the work of R&D having particular characteristics, research suggests that the people involved in R&D projects have special characteristics too. First, they tend not to accept hierarchical control; are less loyal to their organisation than to their profession and they also tend to be chauvinistic about their technical speciality (Gomez et al; Miller, 1986; Von Glinow, 1988). Lack of respect for hierarchy makes them difficult to manage; moreover, since autonomy is important in performing their work, this exacerbates the management challenge (Kim et al., 1999).

Innovation research primarily outlines that R&D team members to possess high intrinsic motivation (Deci et al., 1989), considerable technical expertise (Janz et al., 1997), and a high need for autonomy (Realin, 1985). Consequently, they are expected to have a low need for leadership (de Vries et al., 1999). Thus, one conclusion from this stream of research suggests giving R&D teams as much intellectual autonomy and as little guidance as possible in managing projects.

2.4.2 The Role of the Team Leader in Innovation Projects

Effective leadership makes innovation and thus organisational growth possible (Tushman and O'Reilly, 1997). Getting the balance right between the linear, efficiency that organisations demand and the divergent fluidity that is most conducive for team-level creativity is a delicate matter and would seem to demand very experienced and intuitive leadership (Trott, 2005). Buijs (2007) expresses the role of the leader very vividly:

> Innovation is about coming up with and implementing something new. It is about searching for ideas, exploring ideas, developing ideas, implementing ideas and successfully introducing the ideas (products) into the marketplace. Innovation leadership is about bridging the gap between

dreams and reality, past and future, certainty and risk, concrete and abstract, us ('we love innovation') and them ('they don't want to change at all') and success and failure. And all of these dualities are present at the same time. (p. 204)

Team leaders perform a key role in innovation projects and in facilitating innovation in organisations (Montes et al., 2005). In innovation projects, the role of a team leader is paramount; he or she 'offers a new set of ideas and articulates enough imagination to create a new vision which narrows attention and rallies unity out of diversity' (Cheng and Van De Ven, 1996: p. 610). Team leaders can help coordinate action when members otherwise would not know what to do and this circumstance is likely to arise in the situations of high uncertainty that generally characterise the early phases of NPD research projects. Edmondson (2003) suggests that the literature has downplayed the role of the team leader and has not provided helpful insight into the dynamics of team interaction in intense and uncertain situations which are the contextual hallmarks of innovation teams. Leaders provide enthusiasm and support for creative ideas by protecting new ideas from premature evaluation, recommending new ideas, and recognising and rewarding the generation of new ideas. They also seek access to the resources necessary for creative ideas to be developed. Further, they recognise individuals' contributions to the innovation. Finally, leaders can also contribute to innovation by getting involved and encouraging others to get involved in developing new ideas (Mumford, 2002; Howell and Boies, 2004).

Given the high level of uncertainty combined with an ever increasing level of task complexity that characterise innovation projects, Buijs, (2007) notes that:

CEO's who ask for innovation should be aware that this kind of leadership is rare. It is advisable to search for people who are able to handle this controlled schizophrenia, stimulate them, protect them, believe in them and then let them guide the innovation process. (p. 209)

In the literature on the role of teams generally, the issue of team leadership emerges as 'one of the Big 5' themes (Salas et al., 2005). Paradoxically, the literature on innovation teams, where the context is characterised by high levels of uncertainty, fluidity combined with intense pressure, team leadership has not apparently been adequately recognised as a central factor in the success of those teams. Buijs (2007) observes a gap in the literature in the area of leadership of innovation teams, which he notes 'has hardly been discussed in the innovation literature.' (p.203). This gap in the literature has also been highlighted by Mumford et al, (2002); Edmondson (2009) and Nippa (2006) contends that:

> Comprehensive reviews of the broad research on critical success factors of managing product innovation in most cases do not emphasise leadership or leadership styles explicitly. (p. 2)

Despite this view, I was able to identify in the literature a stream of research that explores the ideal characteristics of the R&D team leader (Table 2.5). The R&D team leader, in many cases, if not synonymous with the innovation team leader, represents a close proxy. Most articles included in Table 2.2 conclude with an inventory of skills that if possessed by the R&D team leader, seem to make their projects generally far more successful.

Table 2.3: Research Papers on R&D Team Leaders, 1980 - 2010

Studies	Context	Empirical or Conceptual	Level of Analysis	Roles/Functions of R&D Team Leader
Maidique, 1980	Innovation Process	Conceptual	Organisation	Technologist; Product Champion; Executive Champion; Sponsorship and Technological Entrepreneur
Souder, 1987	Innovation Process	Conceptual	Individual	Leader, capitalist, exciter, integrator, scout, linchpin, exciter, translator, spotter
Allen et al., 1988	R&D Project	Empirical	Team	Functional Managers Role Project Managers Role
Ancona and Caldwell, 1988	NPD Project	Case Study	Team	Scout, Ambassador, Sentry and Guard
Farris, 1988	R&D Project	Conceptual	Individual	Captain; strategic leader; Organisational leader, technical expert; catalyst; Informal leader, personal developer, climate creator, responsive leader.

Studies	Context	Empirical or Conceptual	Level of Analysis	Roles/Functions of R&D Team Leader
McCall Jnr, 1988	R&D	Conceptual	Team	Technical competence; controlled freedom; leader as metronome; work challenge
Barczac and Wilemon, 1989	NPD Project	Case Study	Team	Communicator, Climate-setter, Planner, Interfacer
Chakrabarti and Hauschildt, 1989	Innovation Process	Conceptual	Organisation	Expert, Champion, Sponsor
Howell and Higgins, 1990	Technological Innovation	Empirical	Organisation	Project champion, technical innovator, business innovator, chief executive, user-champion.
Markham et al., 1991	Technology Innovation	Empirical	Team	Champion; Antagonist
Beatty and Lee, 1992	Technological Change	Case Study	Organisation	Path finding, Problem solving, Implementing
Clark and Wheelwright, 1992	NPD Project	Conceptual	Team	Direct Market Interpreter, Multi-lingual translator, engineering manager, programme manager, concept infuser
Friedman et al. 1992	R&D	Empirical	Individual	Project Management , Personnel supervision, Strategic Planning
Henke et al. 1993	R&D			Task skills (functional and technical) and process skills (interpersonal)
Studies	**Context**	**Empirical or Conceptual**	**Level of Analysis**	**Roles/Functions of R&D Team Leader**
Kessler and Chakrabarti, 1999	R&D	Empirical	Project	Set clear time goals. Find team members with longer organisational tenure; Execute tasks in parallel. Avoid design for manufacture and CAD
Youngbae Kim et al., 1999				

Barczak and Wilemon, 2003	R&D	Case Study	Team	Interpersonal, Project Manager and technical skills.
Thamain, 2003	R&D	Empirical	Project	Creates climate for creativity – a sense of community across the entire organisation. Should be action oriented providing resources, plans and directions for team members Helps identify and diffuse problems at an early stage. Has the ability to engage top management – to get resources, to ensure visibility for the work. And overall support for the innovative projects Facilitate a climate of active participation, minimal dysfunctional conflict and effective communication.

Studies	Context	Empirical or Conceptual	Level of Analysis	Roles/Functions of R&D Team Leader
Filippini et al., 2004	R&D	Empirical	Team	Objective setter; Strategy developer; Clear Communicator; Facilitator of customer and supplier involvement and Project Manager.
Sarin and O'Connor, 2009	R&D	Empirical	Organisation	Motivator, Goal setter, Protector (protecting the team from any attempt to micromanage)
Edmondson and Nembhardt, 2009	R&D	Conceptual	Team	Project management skills; broad perspective; teaming skills; expanded social network; and boundary-spanning skills.

These papers collectively argue that an R&D team leader must both have 'soft', people skills as well as project skills in managing complex, often technical, projects. Some suggest that the team leader should also possess a high level of technical skills so that she can assess the quality of the work being produced. In later papers, as Open Innovation increasingly becomes a feature of R&D processes and context, more emphasis is given to skills like boundary-spanning and networking.

Many researchers argue that employees who are empowered and autonomous exert a higher level of control over their work. This degree of control means that employees feel more comfortable in their role to be innovative in within their own jobs (Thamhain, 1990; Tang, 1999; Zwetsloot, 2001; Nystrom *et al.*, 2002; Amar, 2004; Mostafa, 2005; Muthusamy *et al.*, 2005).

The need for effective leadership could hardly be higher than within NPD teams. This point is made by Mumford et al. (2002) who note that the need for collaboration among team members places a premium on leadership. Strong team leadership includes the ability to direct and coordinate the activities of other team members, assess team performance, assign tasks, develop team knowledge, skills, and abilities, motivate team members, plan and organise, and establish a positive atmosphere (Cannon-Bowers et al., 1995; Hinsz et al., 1997; Marks et al., 2000).

These are all essential elements of leadership in teams. Team leaders also facilitate team problem solving. They provide performance expectations and acceptable interaction patterns and ways of working for members. They synchronise and combine individual team member contributions. Also, they seek and evaluate information that affects team functioning. They clarify team member roles and engage in preparatory meetings and feedback sessions with the team. (Klein et al., 2007; Burke et al., 2006; Stewart and Manz, 1995; Zaccaro, Rittman, and Marks, 2001).

Dolan et al. point out that, in innovation, it is especially necessary to develop a style of 'facilitating' leadership to ensure that the right things happen (2003). They see the essential characteristics as being the capacity to inspire, to articulate a vision and to hold teams of creative individuals together and channel their work. Amabile and Khaire (2008) agree and point out that the leader's job is to map out the stages of innovation and recognise the different skill sets, processes and technologies that are necessary to support each phase. Their simple advice to people managing innovation and creativity; 'Know where you are in the game.' (p. 104)

Given the range and depth of skills required to manage an innovation team and successfully oversee an innovation project, as suggested by the research on leading R&D teams; it seems pertinent to ask whether one individual can possess all those skills or whether the task of managing the project ought to be split and a different leader assigned according to the phase the project is in. Is it plausible to expect one person to be capable of being a visionary, explorer inspirer, inventor, motivator, crisis-manager, coach, expert, boundary-spanner, networker, facilitator, problem-solver, mentor, champion, persuader, ambassador and implementer? Academic research has yet to provide a conclusive answer for this question.

2.4.3 Types of Leadership

Bass (1990, p. 6) reports Napoleon's quip that he would rather have an army of rabbits led by a lion than an army of lions led by a rabbit as indicative of the importance of leadership in teams. According to Eisenhower, 'leadership is the ability to decide what is to be done and then to get others to want to do it.' (quoted in Larson, 1968, p. 21).

Many approaches to the study of leadership exist (see Bass, 1990; Yukl, 2006) but, according to Judge and Piccolo (2004), "transformational–transactional leadership theory dominates current thinking about leadership research" (p. 762). Burns described transactional political leaders as content to work within a framework for the interests of their constituents; he referred to them as opinion-leaders, bargainers, bureaucrats, legislative leaders and executive leaders. Transformational leaders, on the other hand, move to change the framework. He classified them as intellectual leaders, leaders of reform and revolution, heroes and ideologues

Downton (1973) and Burns (1978) presented this paradigm of transformational leadership describing it as a leader who asks followers to transcend their own self-interest for the good of the group, organisation or society; to consider their longer term needs to develop themselves rather than just their needs of the moment. Hence followers are converted into leaders and this is the *transformation* alluded to in the term.

Based on Burns' (1978) influential work, Bass (1985) applied the theory from politics to the firm and organisational context and developed a theory of transformational leadership. In his 'Full Range of Leadership Theory', Bass (1985) described a full spectrum of leadership styles, distinguishing between laissez-faire, transactional leadership and transformational leadership. Laissez-faire, the so-called 'nonleadership factor' (Northouse, 2007, p. 186), is characterised by the absence of leadership. For example, the laissez-faire leader does not set goals, does not provide feedback nor do they provide support to followers in their efforts. Laissez- faire leader behaviour is characterised by avoiding decisions, hesitating to take action, resisting expressing views, delaying responses and being absent when needed (Hinkin and Schriescheim, 2008). Although the data indicate that laissez-faire leadership or nonleadership are negative forms of leadership (Skogstad et al., 2007), academic studies on laissez-faire leadership are more rare than examinations of other types of leadership (Dumdum, Lowe, & Avolio, 2002; Lowe, Kroeck, & Sivasubramaniam, 1996). Hetland (2007) agreed with this assertion suggesting that "laissez-faire leadership behaviour is a destructive leadership behaviour" (p. 80) that warrants much more future research.

Bass (1985) and Yukl (2002) concluded that because most R&D teams are cross functional, that transformational leaders can convince members, through charisma, providing inspiration and acting as a coach and mentor, to look beyond their individual and functional backgrounds and work more creatively. In fact, charismatic leadership is seen as the primary component of transformational leadership in which subordinates are inspired to an enhanced level of performance by the vision and expertise of the leader (Bass, 1985; Pawar and Eastman, 1997; Yukl, 2002). Bass (1995) suggested that the transformational leader also provides intellectual stimulation to subordinates and that, although related to charisma, is a distinct element of the leadership style. Kanungo (1987) saw the transformational leader as one who engages in behaviours counter to the prevailing norms. Keller (2006) suggests that it is quite natural for R&D teams to be positively motivated by intellectual stimulation and that their performance is likely to be helped by a leader who might suggest an alternative way of approaching a problem or a different source of scientific or technical information.

It could be argued that different phases of the innovation process demand different leadership styles. It has been consistently argued that transformational leadership is the preferred style for encouraging creativity and innovation through developing and inspiring followers as well as providing them with intellectually stimulating tasks. This is most critical in the idea generation phase of the innovation process (Hansen and Birkinshaw, 2007). Transformational leadership can often persuade the team members to transcend their own self-interest for a greater, collective cause (Howell and Avolio, 1993). Keller (2006) found precisely this correlation in his longitudinal study in which one hundred and eighteen project teams from five industrial R&D organisations, from a range of industries, were surveyed about the influence of leadership style on their performance. The inspiration and intellectual stimulation effects of transformational leadership was found to be more effective in research projects than in development ones. He notes that research projects 'usually deal with more radical innovations that require originality and the importation of knowledge from outside the project team'. By contrast, the development projects generally focus on incremental innovations and modifications to existing products and more of the scientific and technical knowledge is likely to reside within the team. In development teams, the study found that initiating structure was more effective. Initiating structure refers to a system where the leader defines, directs and structures the roles of activities towards the attainment of the group's goals (Bass, 1990; House and Aditya, 1997; Yukl, 2002). Keller's findings are supportive of a contingency approach to transformational leadership in R&D teams.

A recent paper looking at the impact of transformational leadership on R&D teams has cast doubt on the significance of transformational leadership on the R&D process. Eisenbeiß and Boerner (2010) studied 52 team leaders and 256 team participants in various R&D settings and did not find a correlation between transformational leadership and performance or output. In fact, they found that giving teams autonomy was just as potent as high levels of transformational leadership. This makes it plausible that laissez-faire leadership may have some application in the R&D field. They argue:

> As R&D teams are creative and innovative under low and high levels of transformational leadership, the question arises whether high levels of transformational leadership are necessary at all in the R&D context. Trusting in R&D team members' high intrinsic motivation to innovate and their expert knowledge and thereby protecting their intellectual autonomy thus seem to be as effective as providing high levels of transformational leadership. (Eisenbeiß and Boerner, 2010; p. 369))

Apart from being seen in the leadership literature as an effective catalyst for innovative behaviour and creativity, transformational leadership has also been shown to be the more effective method of recruiting champions to give their support to innovation projects (Howell and Shea, 2001). Innovation champions can make a crucial contribution to innovation projects by promoting their progress through organisational hurdles. Transformational leadership is also commonly linked to change-oriented leadership through which organisations can adapt themselves to a changing environment thus enhancing their capacity and potential to innovate (Yukl, 1999).

In summary, there is much support in the literature for transformational leadership as being especially effective in the area of new product development, although recent studies are suggesting that team members' intrinsic motivation exert just as powerful an influence (as any particular style of leadership) on their creative performance. However, transformational leadership carries the additional benefit of assisting in the recruitment of champions for the team's ideas. Keller (2006) found that transformational leadership is has more positive impact at the front end of innovation projects where the focus is on generating novel ideas. Transactional leadership then emerges as the more effective paradigm once the ideas have been harvested and the focus has switched to their development and implementation or launch. These findings would suggest that when examining innovation projects using the IVC (innovation value chain), that transformational leadership would most effective at the idea generation stage; and possibly at the idea conversion phase. But, transactional leadership would, arguably, be most effective at the diffusion phase when the ideas have been fine-tuned and developed to a state to be ready for introduction.

Support for creativity, risk taking and tolerance of mistakes are norms that are necessary to promote creativity or new ways of doing things in organisations. Teamwork and speed of action are norms that relate directly to implementation (Kessler and Chakrabarti, 1996). Both pairs are required; the first to generate and, the second to implement new approaches. Thus, the results support the idea that two separate component processes (creativity and its implementation) may affect innovation and consequently, innovative teams in organisations. It also further supports the notion of a divide between R&D with the first two factors (support for creativity and risk-taking) impacting more strongly on research and the second two factors (teamwork and speed of action) impacting more strongly on development.

There may be a differential potency of various factors along the innovation process. Cheng and Van de Ven (1996) found that the actions and outcomes experienced by innovation teams tend to exhibit a chaotic pattern during the initial period of innovation development – the research phase, and an orderly, regular pattern during the concluding development period. They described the process as an 'innovation journey' which typically consists of intrapreneurs who, securing support and funding of more senior managers or investors, undertake a sequence of events that aim to create and convert a new idea into an tangible asset or innovation.

Depending on the scope of the innovation, this journey can vary greatly in the number, duration and complexity of events that unfold along the way from the initiation of a developmental project to its implementation or termination. Whatever its scope, this journey is an exploration into the unknown process by which novelty emerges. This process is often characterised as being inherently uncertain, dynamic, and to follow a seemingly random process (Kanter, 1988; Jelinek and Schoonhoven, 1990, Quinn, 1985). Spivey et al. (1997) suggested, 'those professionals who are charged with improving the new product development process may well feel as if they have been asked to bring order out of chaos.' Amabile and Khaire (2008) also acknowledge this so-called chaotic phase of the process; 'Because it's impossible to know in advance what the next big breakthrough will be, you must accept that the discovery phase..... is inherently muddleheaded.' (p. 104)

Cheng and Van de Ven (1996) were concerned that not enough research had focussed upon the early stages of innovation, the generation of initial ideas and the fleshing out of potential developmental platforms of discovery because most researchers had devoted themselves to the more measurable elements of diffusion of innovation (Rogers, 2003) and the implementation of already developed innovation. In their study, they found that actions and outcomes experienced by innovation teams, in the research phase, were not consistent with an iterative, trial and error process; nor did they fit a random pattern – they were, at the research stage, chaotic. The recognition of the element of chaos was originally chronicled over twenty-five years ago by Quinn (1985). In their fieldwork, Cheng and Van de Ven examined two major collaborative pharmaceutical teams who worked for, in one case nine years and in the other twelve, and they argue that the innovation journey begins in chaos and ends in order.

2.5   Summary

In conclusion, innovation performance is a key indicator for most firms. Senior management in most organisations are increasingly seeking proven methods to enhance their company's capability at new product and service development. R&D is the division in many large firms that is responsible for delivering on the firm's innovation goals. R&D departments usually organise into teams and it is teams that develop the innovation around projects and within firms.

Existing models of the innovation process seem inadequate for providing actionable insight for organisations pursuing radical innovation. Radical innovation is articulated as the innovation quest for many organisations. Consensus is emerging that radical innovation, although considerably harder to develop, offers richer opportunities to organisations than steady-state, incremental innovation. It offers higher returns but comes with attendant higher risk. Consequently, researchers are calling for alternative approaches to creating an organisational infrastructure conducive to developing radical innovation (Leifer, 2001; O'Connor and Ayers, 2005; Phillips et al 2006).

The fast pace of technological advance combined with the global nature of knowledge and market creation means that organisations are increasingly engaging in Open Innovation networks (Chesbrough, 2003; 2006). Innovation-centric organisations are supporting inter-firm co-operations and the development of an ecosystem of firms, sharing technologies and trading intellectual property, within a given industry or sector (West et al., 2006).

In managing individual innovation initiatives or portfolios of NPD projects, most large organisations are using some form of the stage gate model. While this provides a helpful managerial framework for new product development, it has attracted criticism (Trott, 2005; Storey and Salaman 2005) principally for constraining the high level of creativity which is a necessary ingredient of innovation projects.

Teams are the organising nucleus for innovation. New product and service development is dependent on organisation members' ability to combine and exchange knowledge (Smith et al., 2005) and hence teams are considered a natural and powerful option for creating and circulating innovative ideas. Teams will be both the source of the creativity and also they will be the mediating instrument that guides the ideas through the organisational innovation process. This innovation journey starts with a raw idea and, ideally, ends when that idea becomes a tangible, profitable innovation asset.

This journey concept argues that innovation should be viewed as an end-to-end process incorporating three distinct elements: Idea Generation; conversion and diffusion (Birkinshaw and Hansen, 2007). Understanding that there are three phases and managing where one is in the process (Amabile and Khaire, 2008) may be a key determinant in successfully managing innovation projects. The innovation value chain is a framework indicating the three stages of a journey that all innovation projects must go through in order to progress from the heartbeat of an idea to the introduction of a new product or service.

Despite their familiarity with the market and dominant technologies, incumbent firms with market leading positions have a poor record in developing radical innovation (Doughtery, 1992). Many factors conspire against larger firms being able to adopt the entrepreneurial agility associated with opportunity recognition and new product development (Markides, 2002). Among the barriers within large firms, the attitude of influential managers to innovation overall is a significant hurdle (Storey and Salaman, 2005) as is the tendency for managers to want to conform and not be associated with risk (Pech, 2002). Some of these barriers are especially prevalent in R&D departments where the work is demanding, time-consuming, resource-intensive but inherently uncertain and risky (Kratzer, 2006)

Strong team leadership includes the ability to direct and coordinate the activities of other team members, assess team performance, assign tasks, develop team knowledge, skills, and abilities, motivate team members, plan and organise, and establish a positive atmosphere (Cannon-Bowers et al., 1995; Hinsz et al., 1997; Marks et al., 2000). Dolan et al. (2003) point out that, in innovation, it is especially necessary to develop a style of 'facilitating' leadership to ensure that the right things happen.

Team climate has been shown to be an important mediator in the development of innovative ideas within organisations.   Risk taking, failure tolerance and support for creativity have been specifically identified with the early phase of the innovation process.

The innovation process is 'inherently muddleheaded' (Amabile, 1999).   It requires a purposeful navigation and management of ideas, projects, organisational resources, teams, processes, customers and deadlines.   As such, one of the crucial decisions senior managers should make in innovation is the selection if team leader for these projects. The role of the leader is critical in managing the progress of ideas from the invention phase through to implementation.

Questions that are not adequately answered in the current literature surround the leadership of innovation teams; a topic which many researchers believe not to have been adequately explored (Mumford, 2002; Nippa, 2006; Buijs, 2007; Eisenbeiß and Boerner, 2010).   Also, although the literature suggests a three phases of innovation projects (Hansen and Birkinshaw, 2007; Roper et al., 2008), there has been little discussion of the differential importance of the three phases.   There are reasonable grounds for expecting that quality of ideas is the key determinant of the success of innovation projects.   If the quest for radical innovation must start with radical ideas, should managers build in a bias in favour of time and resources allocated to the idea generation phase rather than the other two phases?   This is the phase increasingly referred to as the 'Fuzzy

Front End' and is the phase where external linkages play a central role in generating novel ideas. This view would prioritise creativity over implementation in resourcing innovation projects. It would also support management interventions with R&D and/or NPD staff to enhance their creativity.

More specifically, extant literature has established a connection between certain organisational elements and the outcome of radical innovation. Some associations are positive: open innovation networks; teams, transformational leadership; culture; team climate while the literature suggests that other elements have a negative impact on radical innovation; rigid stage gates, closed innovation models, transactional leadership; management culture and aversion to risk.

Informed by this literature, this study will look in depth at the interplay between the various elements and actors in the firm-level innovation process. The objective is to understand the complex relationships that exist between the factors affecting innovation management, throughout the phases of the innovation value chain to allow for a more complex view of innovation in the organisational context.

Chapter Three

Methodology

3.1   Objectives of the Chapter

This chapter has four main objectives: to outline the research objective and questions; to describe the research philosophy; to explain the research design and to describe, in detail, the research process for this study. The table below (Table 3.1) details the specific element of research design for this study. This chapter discusses the various factors which led to these decisions.

Table 3.1: Key Elements of the Research Design for this Study

Philosophy	Interpretive
Approach to Theory	Inductive
Methodology	Case Study
Research Instrument	Semi structured, in-depth interviews
Sample Design	Census of participants in project
Data Collection (Primary sources)	Recorded and typed interview transcripts
Data Collection (Secondary sources)	Company reports, project plans, videotapes of consumer focus groups, briefing materials and presentation slide decks. HR department's 'close-out' report.
Data Analysis	Thematic Coding
Time period	Longitudinal study from 2007-2009
Data Presentation	Writing Up the Data in thesis format
Quality Measures	Trustworthiness • Credibility • Transferability • Dependability

3.2   Research Objective and Research Questions

By definition, innovation requires 'a shift from the norms of average behaviour' (Stevenson and Jarillo, 1990, p. 20). Hence, it is a topic that not merely invites but requires an in-depth evaluation of the people, processes, situations, events and contexts in which it happens (Savage and Black, 1995). Sundstrom and Zika-Viktorrson (2009) acknowledge that although innovation is being studied with ever increasing frequency and intensity, there is still very limited knowledge of how internal (organisational) factors and external factors affect how innovation actually takes place within firms. Van de Ven and Poole (1990; p. 311) hold that 'an appreciation of the temporal sequence of activities in developing and implementing new ideas is fundamental to the management of innovation.'

The objective of this study is to focus on the current practices in the management of innovation within GSK and to evaluate the degree to which the company optimises the organisational elements, which are conducive to the development of radical innovation.

In terms of the study questions, this piece of research is intended to identify how raw, commercial ideas are generated, built and shared in teams; how they are sifted, prioritised and ranked; how some ideas are selected and others are abandoned; how team members network with external experts, lead users and others and bring the learning back in for the benefit of the group; how the ideas are converted from nascent, fluid fragments into clear, concise and testable concepts; how teams can harness their collective creativity and yet still manage the level of efficiency and project management necessary to make things happen; how ideas can be championed effectively within an organisation; how some ideas 'stick' and others don't; how teams should be managed and how leaders should behave in order to support the team to accomplish these goals across the phases of the innovation process.

By answering these questions, through the in depth observation of two innovation teams in GSK, the research should produce useful, and possibly transferable, insights on what precisely are the organisational elements which are most conducive to the successful development, specifically, of radical innovation in R&D settings within large, complex, high-tech, global organisations.

3.3 Describing the Research Philosophy

Research with the objective of exploring, describing, understanding or evaluating topics in innovation management, such as the organisational factors conducive to developing radical innovation, are primarily investigating a social phenomenon (Blaikie, 1993). Authors such as Miles and Huberman (1994), Robson (1993), Yin (1994), Gill and Johnson (1991), Easterby-Smith et al, (1991), Silverman (1999) and Blaikie (1993) show there are various options about how to approach such a social inquiry and that methodological choices are made according to the nature of the questions being addressed and the perspective of the researcher.

Differences in research perspectives are often characterised as a debate between two major and opposing world views or methodological paradigms. The analogy can sometimes extend to characterizing one as 'in the red corner' with the other 'in the blue corner'. (Patton, 1990). Describing these two overarching approaches in the social sciences, Benton and Craib (2001, p. 119) put it: 'there are two basic options; positivism or some form of interpretivism'.

These approaches maintain distinct and different positions on the relationship between ideas and evidence (Miller and Brewer, 2003). The research philosophy chosen will, according to Saunders et al (2007, p. 102), contain important assumptions about the way in which one views the world and these assumptions will underpin the research strategy. They hold this is principally because the research paradigm selected makes a fundamental statement about the author's particular view of the relationship between knowledge and the process by which it is developed or generated.

One key question which will play a role in determining the research approach is whether the social world is seen, by the researcher, as something that is external to those individuals within it, or is it a creation that they play some role in forming? (Bryman, 2004). Thus, two opposite positions or paradigms exist: positivistic (sometimes called scientific or normative) and naturalistic (frequently referred to as constructivist or interpretive). If the world is considered to have just one truth; or a question to have just one answer, then a positivistic approach is taken. If truth, or the answer to the research question, on the other hand, is believed to be constructed from the perceptions of each individual in that social context, and that it may change with different individuals, then the approach is naturalistic or interpretive.

Table 3.2, below sets out what are considered (Bryman, 2004, p. 20) the fundamental differences between the qualitative (interpretive, naturalistic) and quantitative (positivistic) approach in social research:

Table 3.2: Fundamental Differences between Quantitative and Qualitative Research Strategies

	Quantitative	Qualitative
Principal orientation to the role of theory in relation to research	Deductive: testing of theory	Inductive: generating theory
Epistemological Orientation	Natural Science Model, in particular positivism	Interpretivism
Ontological Orientation	Objectivism	Constructionism

Source: Bryman (2004, p. 20).

In comparing qualitative and quantitative approaches, McClintock et al (1979) noted that it is the differences of method that are often emphasised (Fienberg, 1977: 50). Qualitative methods are described as "thick" (Geertz. 1973: p6); they can contain insights and privileged information and: 'can make a major contribution to the development of a meaningful survey design' (Sieber. 1973:

p.1342): they are "holistic" (Rist, 1977, p. 44). In contrast, quantitative approaches can be characterised as "thin" (Geertz, 1973, p.6). "narrow" (Rist, 1977, p. 47), but 'hard and generalisable' (Sieber, 1973: p.1335).

In an essay, titled 'Peaceful Coexistence in Psychology', Taylor (1985, p. 117-120) distinguishes between these two broad philosophical positions in the social sciences. Those that favour the Positivist approach, he refers to as 'correlators'. He sees this approach as relying on what he calls 'brute data' (i.e. data available to the scientist without the need for any interpretation) and is driven by the aim of finding one, singular, unvarying meaning for the data. Taylor contrasts the 'correlators' with what he refers to as the 'interpreters'. These recognise human beings to be self interpreting, whose actions and perspectives are inevitably tied up with how they see the world, their emotions and their values. Taylor characterises 'correlators' and 'interpreters' as extreme points on the social science spectrum; he acknowledges that few investigators would fall entirely into one or other category. Nevertheless, the archetypes, of researchers at either end of the spectrum, are useful to uncover possible differences between the natural and the social sciences.

Rosenberg (1988) summarises the dichotomy by asserting:

> When social scientists choose to employ methods as close to those of natural science as possible, they commit themselves to the position that the question before them is one empirical science can answer. When they spurn such methods, they adopt the contrary view, that the question is different in some crucial way from those addressed in the physical or biological sciences. Neither of these choices has been vindicated by conspicuous enough success to make the choice anything less risky than a gamble. (p. 4)

In the business context, Cooper and Schindler (2008) suggest that what they refer to as the 'controversy' between qualitative and quantitative methods is more manifest in business decisions than elsewhere. They suggest that senior managers don't have confidence in qualitative methods because the sample sizes are generally too small and that results are too susceptible to human error or sample bias. However, they note that 'increasingly, however, managers are returning to these (qualitative) techniques, despite their perceived shortcomings, as quantitative techniques fall short of providing the insights needed to make those ever-more-expensive business decisions' (p.163). Thus, once again making the point, that while quantitative studies offer generaliseability because of their large base of subjects or data points; they may often offer only superficial observations and rarely produce the rich insights that organisations need when developing new products and services.

3.3.1 The Positivist Approach

Traditionally research has been conceived as the creation of true, objective knowledge, following a scientific method. From what appears or is presented as data, facts, the unequivocal imprints of "reality", it is possible to acquire a reasonably adequate basis for empirically grounded conclusions and, as a next step for generalizations and theory building. So, the matter has long been conceived. (Alvesson and Skoldberg, 2000; p. 1)

Positivism embraces the quantitative tradition and is the philosophical stance of the natural sciences. French philosopher, Auguste Comte (1798 – 1857) is generally credited with inventing the term 'Positivism'. Positivism exalts science to the highest form of knowledge; indeed, scientific knowledge is seen as the only genuine form of knowledge. Benton and Craib (2001) suggest that this position can be characterised by a number of basic tenets which include:

• 'The individual mind acquires knowledge exclusively from our sensory experience of the world and our interaction with it.

- Any authentic claim of knowledge must be testable by experience, observation and experiment and this excludes knowledge claims about anything which cannot be observed.

- Scientific laws are statements about general, recurring patterns of experience.

- To explain any phenomenon scientifically is to demonstrate it as a 'scientific law'.

- Scientific objectivity rests on a clear separation of testable, factual statements from subjective, value judgments.' (p. 14)

Bishop (2007) notes that under this paradigm:

Theories are taken to be formal descriptions of facts, observable relationships among facts, and generalisations about facts and their relations. Explanations of observed phenomena are framed as laws and, in turn, laws are thought to be subsumed under more general laws. (p. 14)

Bishop equates positivism with the scientific approach advocated originally by Francis Bacon (1561 - 1626), often hailed as the father of modern scientific method, who held that science starts by making specific observations; it then proceeds to empirical generalisations and tries to uncover predictable and lasting patterns or laws. These observations and generalisations often give rise to the necessity for a new vocabulary of observational terms describing the particular patterns of behaviour of the phenomenon under investigation. Formal language is eventually customised to define the universal laws and principles that emerge.

Positivism in social science can be seen as a natural attempt to put the study of human, social and organisational interaction on a scientific footing by extending the methods and forms of observation, testing and explanation which have held sway in the natural sciences. However, extending the philosophy of the natural sciences into the social sciences continues to be a matter of controversy. Winch (1958) argues that the concepts appropriate to the analysis of social phenomena are demonstrably incompatible with those used in natural science. Similarly, Hayek (1955) contends that modelling the study of social phenomena after the natural sciences has done great harm. Positivism or the scientific approach usually begins with a hypothesis or theory to be proved or disproved. Naturalistic or interpretive approaches differ at the outset because they typically start with an immersion in a situation, allowing the themes to gradually become apparent.

Positivism is a traditional approach to research design in which hypotheses are developed, experiments and surveys (or other research methods) are conducted to prove or falsify the hypothesis. Rosenberg (1995, p. 11) explains that Positivists named their theories 'hypothetico-deductivism' and they held that the history of science is inseparable from a history of progress. This, they describe as a history of increasingly powerful predictions and increasingly precise explanations of how the world works.

The hypothetico-deductive method is generally represented as a logical, linear sequence starting with an a-priori theory with the objective of either confirming or refuting it; see figure 3.1 below.

Figure 3.1: The Process of Deduction

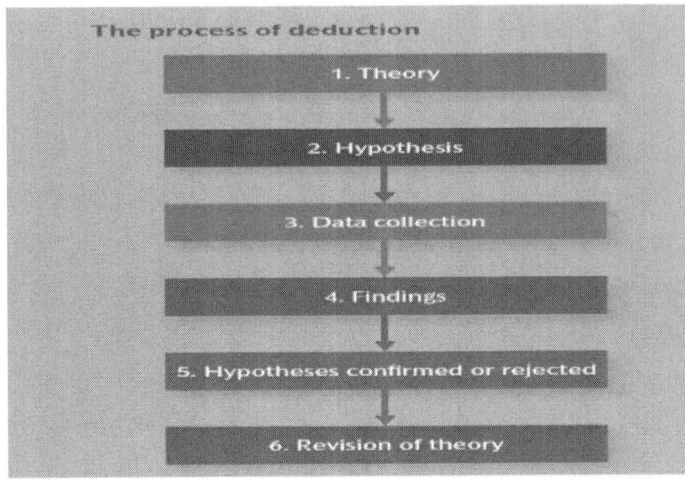

Source: Bryman (2004, p. 9).

Positivist methodologies are associated with quantitative research, which generally focus on causal explanations. In this way, theory precedes data whereas in the inductive approach, the sequence is reversed and theory follows data.

3.3.2 The Interpretive, Inductive Approach

Inductive, interpretivist researchers believe that social phenomena (like organising for radical innovation) cannot be cleanly extracted from context, thus creating complexities for investigations in the social sciences.

Hermeneutics is the name given to the science of interpretation in general. Students of hermeneutics believe that human action is connected or even explained by desires and beliefs which are personal and individual. They hold that methods employed in the social sciences must reflect the distinctive human capacity for language. Hence, measuring physical, human activity in some sphere in search for causal regularities must be considered subordinate to the search for meaning in the social sciences. Bryman (2004, p. 13) holds that at the heart of this clash is a division of emphasis on the explanation of human behaviour that is the principle priority of the positivist approach to the social sciences and the understanding of human behaviour, which is the chief concern of the interpretivist tradition. The interpretivist term reflects the views of the many researchers and writers who have been critical of the application of the scientific model to the study of the social world. Bryman (2004) suggests that they (interpretivists) share a view that the subject matter under scrutiny in the social sciences - people, organisations, human behaviour, and institutions - is fundamentally different from the natural sciences. Therefore the study of the social sciences ought to necessitate the use of different research procedures.

Glaser and Strauss (1967) defined the characteristics of inductive research as: 'the grounding of theory upon data through the process of data-theory interplay, the making of constant comparisons, the asking of theoretically oriented questions, theoretical coding and the development of theory' (Strauss and Corbin, 1994, p. 283).

The qualitative paradigm studies the differences between people (and/or situations) and requires the social scientist, through some level of interpretation, to grasp the subjective meaning of social action (Bryman, 2004). That human action cannot be explained unless it is interpreted is not a new thesis. Plato argued in the Phaedo (99 a-b) that human action can only be understood by interpreting its meaning. One of the principal differences therefore between the two approaches is that while the qualitative, inductive approach observes behaviour, activities and phenomena, it also interprets their significance in the appropriate context.

Induction and deduction are often portrayed as mutually exclusive but in reality the issue is more one of emphasis than dichotomy (Fitzgerald and Howcroft, 1998). Mintzberg has long supported (1979) 'research that is as purely inductive as possible' (p. 584). He notes two essential steps in the inductive process. The first is *Detective Work* where patterns and inconsistencies are tracked down. The second is the creative leap, which involves generalising beyond the data themselves. He describes it as a data-led, exploratory approach involving: 'peripheral vision, poking around relevant places and a good dose of creativity' among other things that 'makes good research and always will, in all fields.'

Yin (2003) contends that the case study approach is essential when elaborate social situations are under scrutiny, because one of its strengths as a research methodology is that it affords a strategy for examining composite, real-life situations. The case study approach successfully manages the countless inter-related elements embedded in real-life situations, which combine to create the phenomenon. Idiographic, is how Bryman (2004) expressed the nature of case study research as its aim is to expound the distinctive elements of the event under investigation, while also attending to contextual features.

In summary, although the two research traditions are often portrayed as a debate between 'hard' (quantitative) and 'soft' (qualitative) research, each approach has considerable strengths. Saunders et al, developed a schematic (Table 3.3) to underscore some of the critical differences in research perspectives between the deductive and inductive approaches.

Table 3.3: Major Differences between Deductive and Inductive Approaches to Research

Deduction Emphasises	Induction Emphasises
Scientific principles.	Gaining an understanding of the meanings humans attach to events.
Moving from theory to data.	A close understanding of the research context.
The need to explain causal relationships between variables.	The collection of qualitative data.
The collection of quantitative data.	A more flexible structure to permit changes of emphasis as the research progresses.
The operationalisation of concepts to ensure clarity of definition.	A realisation that the researcher is part of the research process.
A highly structured approach.	Less concern with the need to generalise.
Researcher independence of what is being researched.	
The necessity to select samples of sufficient size in order to generalise conclusions.	

Source: Saunders, Lewis and Thornhill (2007, p. 120).

In some fields of study, a privileged hegemony has been enjoyed by the 'hard' approach (Orlikowski & Baroudi, 1991, Walsham, 1995), to the extent that 'soft' research will always be seen as somewhat inferior if it is to be judged against the prevailing 'hard' standards. However, Fitzgerald and Howcroft (1998) argue an awareness of both the strengths and weaknesses of the two approaches and an attempt to accommodate them pluralistically leads to a far more complete picture.

3.3.3 Ensuring Quality and High Standards in Social Research

Yin (2003) suggests that for any research study to qualify as 'valid' it should pass certain design tests. Three prominent and essential criteria are generally mentioned in the context of evaluating academic research: reliability; replication; and validity. These are concepts which are very closely aligned with the quantitative, positivist tradition but they also have parallels for the interpretivist paradigm.

Reliability is concerned with the stability of the measures used in the research. This facet of research design can be especially important when researchers are examining concepts which are intangible (like management practices) and may well be emotive; such as poverty, religion and racial prejudice. Reliability is particularly important in quantitative research where the researcher will be concerned that the measures they use are consistent and stable.

Replication is important when researchers choose to re-run research projects and studies of others. There may be many good reasons to try to replicate the work of a previous study. But, for any study to be replicated in a later version the procedures must be very well explained in the original study. That a study is repeatable using precisely the same methodology is sometimes seen as a quality measure for a study.

Another important criterion for evaluating the quality of research is validity. Validity is concerned with the appropriateness and integrity of the conclusions that are generated from any specific piece of research. Validity is generally measured in two ways: internal (or construct validity) and external validity. The notion of construct validity refers to the question of whether the measures devised to record some aspect of a concept or outcome really does reflect the concept it is supposed to describe. This question of whether the measures employed really represent the concepts they are supposed to be scoring is a fundamental one.

Internal validity is also concerned with the accuracy of the markers used for causality within a study. It relates to whether a conclusion that attributes a causal relationship to two or more variables is actually correct and that the relationship so attributed is not capable of any alternative explanation. Internal validity measures just how confident we can be that the relationship attributed to the dependent and independent variables in the study is watertight.

External validity relates to the question of whether the results of the study can be generalised beyond and outside the specific research context. Bryman (2004, p. 29) notes that these measures of reliability and validity are 'most obviously a concern for quantitative research'. This is not to say that they are not a concern for qualitative research but to acknowledge that these measures are specific to areas of causal connections which are not always priorities for the interpretive tradition. He notes that 'grounding these measures and criteria in quantitative research makes them inapplicable or inappropriate for social research'. It is important to note (Yin, 2003, p. 37) that generaliseability in the context of a case study does not refer to statistical generaliseability but rather analytical generaliseability where 'the investigator is striving to generalise a particular set of results to some broader theory.' Also, contribution to generaliseability can be attributed to the extent that the research findings replicate extant theoretical propositions or other empirical findings in the field.

Lincoln and Guba (1985), in pursuit of alternative ways of evaluating qualitative research, propose *trustworthiness* as an overarching criterion to evaluate the quality of a research study. Trustworthiness, in their view, involves establishing four key criteria: Credibility - confidence in the 'truth' of the findings; transferability - showing that the findings have some level of applicability in other contexts; dependability - showing that the findings are consistent and could be repeated; and confirmability - a degree of neutrality or the extent to which the findings of a study are shaped by the respondents and not researcher bias, motivation, or interest. Many researchers endorse the concepts credibility, dependability and transferability, which have been used to describe critical components of trustworthiness (Guba, 1981; Patton, 1987; Polit and Hungler, 1999; Berg and Welander Hansson, 2000).

3.3.4 How is Innovation Approached within the Current Literature?

The top-ranked journal dealing with new product development and innovation (Rated A, according to the Harzing Index) is the International Journal of Product Innovation Management (JPIM). This is the official journal of the Product Development and Management Association (PDMA). In 2007, to celebrate twenty years of publication, the JPIM published a number of review articles evaluating both how the journal had been influenced over the two decades of its existence and how it had impacted on other publications. These articles also audited and classified the methodological perspectives of the articles it had published over the two decades. One conclusion (Biemans et al, 2007) was that over the twenty years, the dominant research method used was quantitative; specifically, a cross-sectional, large-sample survey, and the focus most usually being at the level of the firm. The surveys were predominantly administered by mail with some telephone and email questionnaires being introduced more recently. This demonstrates that the JPIM, a key organ of opinion in new product development, leans towards positivist, survey data.

Gua (2005), in a similar JPIM review article and commenting on this phenomenon, holds that in this pursuit of generaliseability and reliability, scholars may depend heavily on a limited palette of survey methods and 'sophisticated software to treat their well-structured data. This reliance risks unintentionally leading researchers to choose the studies that are tightly designed, exactingly executed and straightforwardly measured' (p.255). Like Daniels (1991) and Inkpen and Beamish (1994), Guo believes that the pursuit of statistical validity may lead scholars to continue to examine the same phenomenon repeatedly with a relatively low level of data. There is, he says, a lack of diversification of research methods within the area of NPD research. Interestingly, a fact welcomed by Biemans et al. (2007); case study methods have almost doubled in popularity over the lifetime of the journal (rising from 8% in 1984 to over 14% in 2003) which is an illustration of the increasing plurality of methods both being published in the journal and being undertaken in the field.

The design of this research study takes a different approach to the prevailing, positivist orthodoxy in the innovation field. In that sense, it responds to the call for a more pluralistic approach to the study of innovation. Because the focus of the study was to understand how innovation happens in large organisations, a qualitative approach was preferred. Moreover, the elements that were required to be probed in this study are complex human emotions, skills, responses, creativity, team-working and motivations.

3.4   Case Study and Case Analysis

In terms of epistemology, this study is a qualitative inquiry. Having reviewed the literature; reflected on the practice of innovation in companies and performed the empirical work for this study, a case study emerged as the ideal mechanism through which to attempt to answer the research question.

Denzin and Lincoln, (2000, p.190) point out, that as one begins the practical activity of generating and evaluating data to answer questions about the meaning of what others are doing and saying and then synthesising that insight into a published work:

> One inevitably takes up "theoretical" concerns about what constitutes knowledge and how it is to be justified...In sum, acting and thinking, practice and theory, are linked in a continuous process of critical reflection and transformation.

Case studies are appropriate when conducting exploratory research on complex social phenomena in their real life contexts (Eisenhardt, 1989; Yin, 1994). Additionally, as well as allowing researchers to observe formal, measurable elements of a process, they also reveal the informal elements such as social interactions which can be extremely important (Hartley, 1994). R&D managers have been found to be positively inclined to case study research (Gassmann, 1990).

Govindarajan and Trimble (2010; p xiii) favour case studies as the best way to build knowledge about innovation:

> The only way to study the management of innovation initiatives is to compile in-depth, multiyear case studies. Doing so is time consuming and expensive. It requires extensive interviewing, followed by the meticulous process of synthesising hundreds of pages of interview transcripts and archived socuments into meaningful narratives. This work requires access through unique partnerships with corporations, and corporations are generally willing to partner with only the top academic institutions.

Yin (2009) defines a case study as: 'an empirical enquiry that investigates a contemporary phenomenon in depth and within its real life context, especially when the boundaries between phenomenon and context are not clearly evident.' Yin further asserts that a case study is especially appropriate when the objective is to understand the workings of a real-life phenomenon in depth including contextual factors relevant to the study (Yin and Davis, 2007).

Yin considers case study to be better able to rationalise the causal relationships in real-life events than empirical strategies. Case study methodology also has the capacity to vividly describe the situation under investigation, providing important contextual detail. It allows issues to naturally come to the surface, and in situations where the results or effects are indistinct, the case study methodology provides space to explore.

For this project, a multiple (twin), longitudinal case study, based primarily on in-depth interviews (Eisenhardt, 1989; Miles and Huberman, 1994; Stake, 1994; Cresswell, 1997) has been selected. In-depth interviews capture an individual's perception of events and behaviours after they occur, allowing important issues to come to the surface. In-depth interviews in a twin or multiple case study have the highest potential to provide deep contextual understanding of the critical factors involved in the innovation process (Eisenhardt, 1989).

Mason (1996) holds that such interviewing is recommended when the evidence being sought is situational, contextual and interactional. Zorn (2001) also suggests the most useful interview format for conducting qualitative research is often "semi-structured". Being 'semi-structured' places the interview at the centre of a range between; highly-structured, as is the case of an interview that consists entirely of closed-ended questions, and unstructured, such that the interviewee is simply given the opportunity to talk freely about whatever comes up. Semi-structured interviews offer topics and questions to the interviewee, but are carefully designed to elicit the interviewee's ideas and opinions on the topic of interest, as opposed to leading the interviewee toward preconceived choices. To be effective, they rely on the interviewer following up with probes to get in-depth information on topics of interest. Mason (1996) refers to this methodology as 'qualitative interviewing', an in-depth, semi-structured form of interviewing. She sees them as generally characterised by:

- A relatively informal style with the appearance of a conversation or discussion rather than a formal question and answer format.
- A thematic, topic-centred approach where the interviewer does not have a set of pre-formatted questions, but rather has a set of themes and topics to explore.
- The assumption that data are generated through the interaction because either the interviewee or the data itself are data sources.

Additional benefits of the case study analysis approach are that it accepts small sample numbers, allows for creative use of alternative data sources (of which there are several in this case study), places emphasis on historical outcomes as the key area to be examined and provides the ability to identify the causal mechanism that underlies the outcomes (Easton, 1995). The case study can also reflect and be sensitive to the context within which a phenomenon occurs and to the temporal dimension through which events unfold (Li, 1996).

3.4.1 Data Analysis: Coding the Data, Linking Concepts and Data

Ezzy (2002, p. 60) notes that data analysis in most qualitative research begins during data collection and that this is consistent with the dialectic theory and interplay between theory and data.

Figure 3.2: Relationship between Analysis and Collection

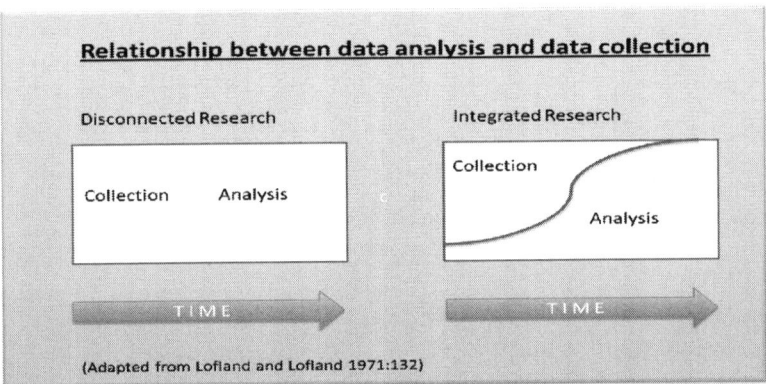

Source: Adapted from Lofland and Lofland, 1971, p. 132).

He notes that theoretical and practical questions are shaped and reshaped as data begins to be gathered. This was the case with this study as, apart from the pilot, the interviews with the team leaders preceded the other ISF interviews and yielded some useful insights that otherwise might have been lost to the study. Many theorists see data collection and analysis as interrelated processes (Glaser and Strauss, 1967; Becker, 1971; Strauss and Corbin, 1991) where they assert that the data analysis begins when the very first piece of data

is collected.   This is particularly true in the case of inductive research where
Ezzy (2002, p. 620) points out:

> Examining data right from the beginning of data collection for "cues" is
> what makes grounded theory "grounded".   It is also the foundation of
> inductive theory building.

Coffee and Atkinson (1996, p. 26) note that most analyses of qualitative data
begin with the process of coding.   They stress that although coding is almost
invariably a part of the analysis of the data; coding is not a substitute for
analysis.   Coding is the process of reviewing the data and allowing the key
themes and concepts to emerge.

The purpose of parsing, analysing and coding data is to allow theories and concepts to emerge from themes embedded in the raw data which can be woven or reconstructed into a theory grounded in the fieldwork; as Glaser states:

> The purpose of coding is to move from having a set of data to developing a theory by fracturing the data, then conceptually grouping it into codes that then become the theory which explains what is happening in the data. (1978, p. 55).

In short, coding is the process of finding out what the data are all about. (Charmaz, 1995, p. 37).

3.5   Research Design

The research design is based around the longitudinal analysis of a double, or twin, case study in one specific firm over the course of three years. The company involved is the second largest pharmaceutical firm in the world and one of the world's top ten R&D spenders. Moreover, the initiative to be studied is unique within the company and is so singular in its design as to be likely to be unique in any setting. The cases to study were chosen after extensive discussions with the company's head of R&D and head of Innovation, head of Investor Relations as well as with one of the company's (Global marketing) Presidents.

The corporate context in which this research was conducted will be extensively described in Chapter Four. The company's overall, published business strategy cites innovation as its key route to and priority for growth. A number of initiatives were extant at the time in pursuit of new product and service development ideas to drive growth in the business. While the ISF was not the only, or even the primary, innovation effort, it was seen as the most experimental and entrepreneurial innovation project that the R&D group had ever inaugurated. As part of the preparation for this thesis, a number of preliminary interviews took place with key stakeholders within GSK to get some insight on which programmes were most appropriate for a study of this nature. The company's senior leaders including the President of R&D, his counterpart in the commercial organisation; the President of the Future Group as well as the company's Global Head of Innovation were interviewed. All recommended this initiative, called Innovation Sans Frontiers, which was then underway within the R&D organisation. This internal innovation initiative was ideally configured as an embedded, twin case study. Gerring (2007, p. 37) defines such an approach as an 'intensive study of a single unit or a small number of units (cases), for the purpose of understanding a larger class of similar units (a population of cases).' This is perfectly aligned with the objective of using the leanings derived from the ISF project to help shed some light on which practices and processes may help accelerate innovation more generally in organisations and, more specifically, in teams.

3.5.1 The Right Case to Study - Why ISF is a Revelatory Case

The Innovation San Frontiers (ISF) project within GSK was ideally constituted to yield fresh, relevant, possibly original and hopefully, transferable insights. It was deliberately set up with a novel framework; certainly one that was unique for GSK. The project involved pitting two equal (in theory) teams against each other to come up with radical new, business ideas capable of being launched, by GSK, within 3-5 years.

Consistent with the processes associated with drug discovery, this twin team approach was almost analogous to a clinical trial where there is a control group but neither side knows which is the placebo and which is the active group. In this case, they were both 'active' groups with neither designated as the 'control' but nonetheless the structure of the ISF initiative makes comparisons possible that could not have been achieved with a more conventional project or team design. The teams were 'created equal' insofar as there were roughly the same amount of people on each team; both teams had an equivalent level of experience within its ranks; both teams had a nucleus of highly qualified and expert PhD scientists within it. The teams were also both working to exactly the same brief, with the same deliverables, the same budgets and resources and the same deadline.

These factors mean that their experiences can be examined at a number of levels for insights into how innovation happens in a large, complex, R&D intensive, pharma company. These factors also make the ISF a case study that falls into the category described by Yin (2009) as 'revelatory' because the case provides an opportunity to look behind the curtain of a large, global, R&D intensive corporation into a specific, unique, situation or context which would not ordinarily be allowed to be the subject of external or academic scrutiny.

The case study was conducted using a longitudinal series of interviews. The approach was inductive and the objective was to allow the participants tell the story of their experience with the project without having to steer them towards any a priori agenda or hypothesis. This research elicited the narrative of what happened over the nine months of the project, how it happened, who did what, how the team collaborated to leverage their collective skills and what results did they achieve. Through their stories, it was hoped that common themes might emerge. It was likely that similar experiences would give rise to common issues and themes which could then be analysed in more detail. Ultimately, the objective was to develop some guidelines which might improve both the experience and the outcomes of similar projects and initiatives in the future.

Members' experience of the ISF project; the briefing, the work expended on generating ideas, following-up technologies, creating prototypes, testing them with potential end-users, establishing a potential value for these raw ideas – and finally, presenting their findings to senior management, were likely to be deep, personal and complex. The research methodology used in this research, therefore, needed to be capable of handling this complexity.

Mason (p. 40) points out that:

What you want to know may be rather complex or may not be clearly formulated in your interviewees' minds in a way that they can simply articulate in response to a short, standard question.

As Gillham describes:

The use of 'open' questions doesn't mean that you have no control over the way the interviewee responds. Indeed, your (unobtrusive) control is essential if you are going to achieve your research aims, i.e. you need to 'steer' for the direction and also ensure that key points or topics are covered. (2000, p. 45).

3.5.2 Sample Design

One of the advantages of studying a customised, time-bounded, internal, organisational experiment like ISF is that the number of personnel involved was limited and this allowed for interviewing the entire target population. A census approach was feasible and therefore the sample design required the assembly of the lists of team members on both teams along with a list of the SLT members who had been involved in designing, recommending the staffing and judging the outputs of the group. Interviews were planned for all members of both ISF teams and with all members of the SLT at the time of the project.

3.6 Describing the Research Process

3.6.1 Fieldwork Phase 1: Deciding on the Right Case to Study

The fieldwork process described in the following pages took place over a period of over two years from March 2007 to July 2009. The first phase in the process was a series of exploratory interviews with the senior leaders in GSK. Meetings with the President of R&D (Dr Ken James) and with his counterpart in the commercial organisation (Tim Wright) were set-up to discuss the research proposal and, with them, to review possible GSK innovation programmes and initiatives which might be suitable for the type of study being proposed.

They also wanted to identify the innovation programme or initiative within the organisation which would be most useful for them to spotlight in this type of in-depth research. These initial conversations were widened to include the company's Head of Innovation (Donna Sturgess). Also involved in these early discussions was the company's Global Head of Organisational Development (Dr Sandy Lionetti). It was during these discussions that the ISF programme, which was then just approaching completion, was first mentioned.

The ISF initiative was a supplementary innovation programme to the company's already extensive, established innovation activities (see Chapter Four) and consequently, there was a strong interest in measuring its effectiveness both in terms of innovation outputs (or ideas for products and services GSK might launch in the future) as well as overall project design. The company was interested to know whether there was merit in doing more of these initiatives and also what they might learn out of the structure, processes and outcomes of this version so that they could improve any future iterations of this type of programme.

The ISF project suited this type of in-depth study for a number of reasons:

123

1) It was a high-level innovation effort supported by senior management

2) It had interesting and unique (for GSK) elements of experiment design insofar as it pitted two teams in competition against each other.

3) It had an international dimension with the two teams located separately

4) It had clear and explicit objectives for the teams

5) It offered unprecedented access to highly qualified and skilled R&D people to conduct the study.

These discussions also helped bring to the surface some of the concerns of senior management about the company's competency and approach to innovation. These initial discussions, at the top (functional president) level of the company, were semi-structured but, essentially exploratory in nature. The sequence of interviews and the structure of the entire project is provided in Figure 3.3.

Figure 3.3: The Research Process

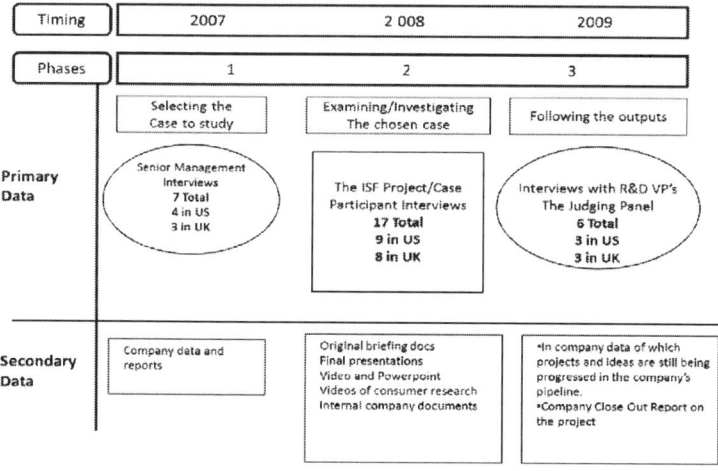

The first phase of the research process identified the ISF project and provided some of the organisational context that surrounded it. These interviews provided rich insights into what senior management were attempting to do with this project. These five interviews were conducted between March and September 2007 and they lasted between forty-five minutes and ninety minutes each. This phase both identified the right project to study and gave the research the imprimatur of senior management to proceed.

As mentioned above, these interviews involved the function heads of the entire organisation; including the global presidents of innovation, marketing and R&D. The project they identified, and recommended to be the subject of this study, was being managed by the R&D president and his team of direct reports. This group of R&D senior managers (all at the level of Vice President within the company) were called the Senior Leadership team (SLT).

In setting up the ISF project (which will be described in detail in Chapter 5), each of the SLT members had been asked to nominate possible members to staff the ISF teams. There were to be two teams with roughly a dozen R&D people in each. R&D was located in two major sites; Weybridge in Surrey and Parsippany in New Jersey. For the ISF initiative, one team would be based in the UK with the other based in the US. The people who were nominated to join the ISF initiative all had to have shown some evidence of an above average level of creativity to justify their selection.

The SLT had three significant and separate roles in the management of the ISF project: they designed the overall organisational element of it and set the aims and objectives of the initiative; they nominated the people who would be working on the ISF project and they judged the quality of the work produced by the ISF team and they decided which (if any) of the resulting ideas would be taken into the overall R&D pipeline.

3.6.2 Pilot ISF Interviews

In keeping with good practice, a preliminary set of 6 (ISF-participant) interviews was conducted in order to pilot and pre-test the topic guide that had been developed for the interviews. These were conducted in early February 2008. Included within this pilot group were interviewees from both (US and UK) teams and one of the team leaders was also included. This pilot group of interviews helped in providing context for the rest of the fieldwork.

3.6.3 Fieldwork Phase 2: Interviewing the ISF Participants

The next phase involved interviewing the ISF team leaders and team members. Seventeen interviews were conducted with the ISF teams between February 2008 and June 2009. There had been 25 members over the entire project but some had left GSK either during the project itself or very shortly afterwards. These interviews, because they explored the experiences and reflections participants of the project, were longer in duration (averaging 1h 45mins) and were recorded and transcribed with many of the transcripts running to over 30 pages in length. The details of the participants and the times and dates of the interviews are recorded here in Table 3.4.

Table 3.4: The Interview Schedule

Team Senior Management	Name	GSK Role/Position	Date of Interview
1	Donna Sturgess	Global Head of Innovation	March 12th 2007
2	Tim Wright	President the Future Group	April 3rd, 2007
3	Dr Sandy Lionnetti	Director of Organisational Development	April 8th, 2007
4	Robert Wolf	SVP HR Global	September 10th, 2007
5	Peter Kirkby	VP Marketing Excellence	September 17th, 2007
6	Gary Davies	Director of Investor Relations	September 24th, 2007
7	Dr Ken James	President of R&D	October 1st, 2007
Team US			
8	Scott Coapman	Team Leader ISF US NPD Project Leader	February 6th, 2008
9	Li-Lan Chen	NPD Scientist	February 6th, 2008
10	Ayyappa Chaturvedula	NPD Scientist	April 28th, 2009
11	Rajesh Mishra	Medical Director	April 17th, 2009
12	Michael Buch	NPR Director	February 18th, 2008
13	Susan Schwartz	Medical Director	April 22nd, 2009
14	Prasad Adusumilli	NPR Scientist	March 30th, 2009
15	Brenda Schuler	NPD Scientist	June 1st, 2009
16	Frank Deng	NPR Scientist	January 17th, 2008

Team UK			
17	Nigel Grist	Team Leader UK	February 18th, 2008
18	Stuart Smith	VP Dental Care	February 8th, 2008
19	Alex Stovell	NPR Scientist	February 5th, 2008
20	David Uruquart	NPR Scientist	February 8th, 2008
21	Ashley Barlow	NPR Scientist	April 9th, 2009
22	David Parker	NPR Scientist	March 25th, 2009
23	Jonathon Creeth	NPR Scientist	April 7th, 2009
24	Peter Frost	NPR Scientist	April 20th, 2009
R&D Senior Leadership Team (SLT)	Name	GSK Role/Position	Date of Interview
25	Stanley Lech	VP, R&D and Innovation, GSK Consumer Healthcare	March 21st, 2008
26	Teresa Layer	R&D VP Sensodyne	October 1st, 2008
27	Kenneth Strahs	R&D VP Smoking Control	September 9th, 2008
28	Geoff Clarke	R&D VP Panadol	May 7th, 2009
29	Brendan Marken	VP NPR	June 2nd, 2009
30	Simon Gunson	VP NPD	July 30th, 2009

3.6.4 Phase 3: Interviewing the Judging Panel

The third, and final, phase of the research involved six interviews with the senior leadership team in R&D. This group had the task of reviewing the ISF work, evaluating all the novel ideas and assessing their potential within the organisation and deciding which projects were going to be progressed after the ISF project concluded. As the ISF project was created to have a competitive element, it was important to ensure the voice of the 'judges' was captured in the analysis. The table above (Table 3.4) shows the timing and number of interviews in which the primary data was captured. Table 3.5 below enumerates the number of pieces of secondary data that also featured in the overall analysis. Broadly, the initial scoping interviews could be referred to as 'Phase 1' and took place towards the end of 2007. The bulk of the work and the key interviews in the ISF teams were undertaken in 2008 and are described here as 'Phase 2'. Finally, the interviews with the SLT of the R&D group took place last and are described as 'Phase 3' and took place in 2009.

Table 3.5: Additional Material: Secondary Material

• **SLT Briefing Documents for ISF Project – PowerPoint Presentation** The initial briefing deck of 6 slides that were developed and used to brief the ISF teams both in the UK and US.
• **US Team Final Presentation to the SLT– PowerPoint** The US team's final presentation including all the concepts they presented at the final presentation.
• **US Team Microsoft Project Plan for the Initiative** US team's project plan which was created at the outset of the project and updated throughout and forms part of the analysis and helps corroborate the timings quoted within the fieldwork section of this report.
• **US Team Consumer Research for the ideas they presented** Includes unedited video footage of a number of focus groups (~40 hours of footage) in which the ideas were tested with consumers. Also included BuzzBack first stage volumetric data on the ideas.
• **The UK Team's Final Presentation to the SLT - PowerPoint** UK Team Presentation including PowerPoint slides showing concepts and context for the ideas presented to the SLT at the final presentation.
• **Internal Company Magazine Describing the Outcome of the Initiative** The R&D Organisation has an internal magazine in which this project was written up.
• **HR Project 'Close-Out' Report** A junior HR executive, based in GSK Head Office, carried out what they described as a 'cursory review and topline recommendations' on the project after it had come to an end.
• **US Team Close Out Report** The US team leader got his team to run a post-project review after the final presentation and their feedback on the experience of being in the team and observations on how it was run were captured on Think-Tank software.

Around the time of the decision to use this innovation initiative as the focus of the research study, the ISF project was already nearing its completion. The project was due to conclude with a presentation of each team's ideas to the R&D Senior Leadership Team (SLT) in New Jersey in July 2007. The UK team was flying over to present their material for the first half of the day and the US team would have the afternoon of the same day to present their top ideas. As part of the research project, this researcher requested and was granted permission to attend this final presentation of the ISF ideas.

3.6.5 Timing of the Interviews

For the initial interviews, in which the ISF project was scoped out and identified as an appropriate one for this study, most of the meetings were face to face and extensive notes were taken and subsequently typed up. But these meetings with the senior presidents and function heads in GSK were generally shorter and more open and exploratory than the subsequent substantive and formal ISF interviews.

Interviews with ISF members took place at least six months after their final presentation in New Jersey (July 2007). These interviews began with an initial telephone call to contact the two team leaders and to seek their permission to conduct a series of interviews both with them and with the members of their teams. Once this permission was received, all interviewees were asked to sign an informed consent form as part of the Dublin City University (DCU) research protocol. The team members were then contacted individually and the same process was entered into with them; interviews were scheduled and they signed and returned a consent form and these forms were lodged with the academic supervisor in DCU. In all cases, prospective interviewees were contacted at least one month in advance of the proposed date of the interview to ascertain their willingness to be interviewed. There was an enthusiastic response to the requests for interviews; nobody refused to be interviewed. In these preparatory calls, candidate interviewees were given an outline indication of the area upon which the research was to be focussed and they were told roughly how much time to set aside for the interview. In all cases, the mention of the ISF project was sufficient prior information for them to prepare for the interview and this was helpful as some of the interviews were carried out some time after the project had come to an end and it was useful that the interviewees had looked at their notes beforehand. Most of the ISF team interviews took place on the telephone and were professionally recorded and transcribed by a professional organisation (Intercall) who provide this as a corporate service. The procedure required both interviewer and interviewee dial into a pre-assigned number, a person from the service provider would host the call and would introduce themselves and make sure that the interviewee was comfortable being recorded in this way and then the Intercall person would 'go on mute', record the call and type it up from the recording. A transcript was provided some three days later.

This approach had as its objective that the interviewees would simply relate their experiences of the initiative. The opening question; 'tell me about your experiences of the ISF team and project'; invited them to give a comprehensive narrative or description of their experience in this context. The only other 'structure' to the research design was that a topic-guide had been prepared with some bullet points with which to prompt the respondent if they did not cover this element in their own account of the experience.

3.6.6 Identifying the Interview Themes

In the case of this study, all the interview transcripts were printed off and assembled, some thirty in all; with some individual transcripts stretching to over thirty typed pages in length. In the first instance the transcripts were read without coding to give an overview of the general direction or flow of the interviews, and to assess the tone and context for any significant or unusual references, comments or patterns. This first reading was done without making any notes on the transcripts. On the second reading, some notes were made about the themes which were recurring most often. On the third reading, an attempt was made to classify each transcript according to the issues that seemed to dominate each specific interview. Interviews that had common groups of similar themes were placed together. Broadly speaking, and somewhat predictably, the transcripts and the themes that linked them seemed to cluster and divide naturally into three overall categories: the US Interviews; the UK Interviews and the Senior Management Interviews

3.6.7 Coding the Data

Coffey and Atkinson (1996) point out that all researchers need to be able to organise, manage and retrieve the most important and meaningful bits of their data and the usual way to do this is by assigning tags, labels or codes to the data. The identification of the relevant and appropriate concepts that rose to the surface in the interviews is important analytical work. As Seidel and Keller (1995) note,

> Codes represent the decisive link between the original "raw data", that is, the textual material such as interview transcripts or fieldnotes, on the one hand and the researcher's theoretical concepts on the other. (p.52)

In doing the coding, the Miles and Huberman (1994) approach was adopted, in which they describe codes as:

> Tags or labels for assigning units of meaning to the descriptive or inferential information compiled during a study. Codes usually are attached to "chunks" of varying size – words, phrases, sentences or whole paragraphs, connected or unconnected to a specific setting. They can take the forms of a straightforward category label or a more complex one (metaphor). (p.56)

The initial identification of concepts, topics or themes is often referred to as 'open coding' (Ezzy, 2002, p. 87). Having created a number of themes that occurred in the raw fieldwork, the next step was to develop a table in which these phrases were posted. These themes were referenced in words, sentences and paragraphs. It is worth noting that all interviews were conducted one-to-one and took place at least three months prior to the commencement of coding and therefore there was no possibility of group think or research bias coming from any interviewee seeing or knowing what the codes are. These codes and themes helped to organise and to retrieve the data. Again, Miles and Huberman (1994) make the point:

> The organising part will entail some system for categorising the various chunks, so the researcher can quickly find, pull out and cluster the segments relating to a particular research question, hypothesis, construct or theme. (p. 57)

This coding in which data are assigned to broad categories at a general level and thereby reduced to manageable proportions can be thought of as a simple form of content analysis (Krippendorf, 1980). This part of the process sorts the data into a straightforward concept schema and then involves indexing the interviews against the codes and retrieving the relevant quotes and placing them appropriately. As Coffey and Atkinson note (1996, p. 29):

> Such coding and retrieving can be implemented in a variety of manual styles. Texts can be marked up physically with marginal keywords or code words, different colours can be used to mark or highlight the texts and index cards can be used to cross-reference instances to numbered pages or paragraphs in the data.

This process is not intended to restrict or limit the interpretations or dimensions of the data; it is not to simplify the data but to allow them be opened up and interrogated further including the attempt to look for further meaning or insights within the data. The creation of overall themes in the data and the continual examination of the raw data against these themes creates a powerful dialogue between the researcher and the data.

The next step in the coding is know as 'axial coding', so described by Strauss and Corbin (1990, p. 97) and this involves specifying a specific category or phenomenon in terms of the conditions that give rise to it and its impact on the respondents. Ezzy (2002, p. 90) notes that the aim of axial coding is to integrate codes around the axes of central categories. The formal process originally set out by Strauss and Corbin (1990) suggests that axial coding should focus on four specific dimensions: context, strategy, processes and consequences.

Table 3.6: Coding in Thematic Analysis

Approaches to Coding in Thematic Analysis
Open Coding
Explore the data
Identify the units of analysis
Code for meanings, feelings and actions
Make metaphors for data
Experiment with codes
Identify the properties of codes
Axial Coding
Explore the codes
Identify the relationship between codes
Compare codes with pre-existing theory
Selective Coding
Identify the central story in the analysis
Examine the relationship between the core code and other codes
Compare the coding scheme with pre-existing theory

Adapted from Ezzy (2002, p. 93).

3.6.8 Writing up the Data

While writing signals the final phase of the research process, it is still a crucial stage within that process. Ezzy (2002, p. 138) notes that for many people discovery occurs in the process of writing as much as it does anywhere else in the process. He goes further to say that writing is about creating results just as much as it is about reporting them. He postulates that qualitative data analysis is an interpretive task where, he says: 'interpretations are not found – rather they are made, actively constructed through social processes.'

Richardson (2000, p. 923) considers writing:

> As a method of inquiry....a way of finding out about yourself and your topic.....writing is not just a mopping up activity at the end of a research project. Writing is a way of knowing –a method of discovery and analysis.

Many writers urge researchers to take the time to reflect on their data. Time 'is essential for the gestation of ideas' (Garrett 1998, p. 29). Using the metaphor of gestation implies that this is a process that both requires and benefits from spending an appropriate amount of time on it. Ezzy (2002, p. 141) agrees

> Interpretations are not discovered solely by following the correct method...rather they are nurtured and discovered through a difficult process not unlike pregnancy and labour. Writing is the moment when the ideas are given concrete form.

Based on the coding and thematic analysis process in this research, four separate themes were identified. These are provided in the table below and will be discussed in detail in Chapter 7. Ezzy (2002) notes that researchers should search for the one 'central story' in the analysis and in this piece of research; the central story or overarching theme is that of team leadership. Team leadership in this context is an issue that is central to the entire ISF story and membership experience.

3.6.9 Themes Emerging from Coding and Thematic Analysis

Figure 3.4: Major Themes that Emerged in the Coding Phase

Key Themes to Emerge from Coding and Thematic Analysis

3.6.10 The Challenge of Insider Research

Participant observation 'involves a level of immersion that allows the researcher to be able to intellectualize what is seen and write about it convincingly' (Tian, 2010. p83). To develop a strong familiarity with the business issues being studied, participant observers must untangle various layers of meaning by engaging in a level of interaction that allows them to test their insights about a setting. Ultimately, participant case study research is an inductive process, whose data is produced by repeated and prolonged contact between a researcher and informant, often with considerable mutual involvement in the personal lives, or occupational contexts of native participants (Tian , 2010).

Participant observation is an approach grounded in a commitment to the first hand experience (Oliver and Eales, 2008) and exploration of a particular social or cultural setting (Atkinson et al., 2001; Kemmis and McTaggart, 2005) with the researcher either becoming actively involved (or having already been involved) in the situation being studied, participating overtly or covertly in people's lives for an extended period of time (Hammersley and Atkinson, 1995; Denzin and Lincoln, 2005).

Through this process, the observer is able to interpret the meanings and experience the interactions of people from the role of an insider (Jorgensen, 1989), enabling the researcher to place specific encounters, events and understandings into a fuller, more meaningful context (Tedlock, 2000).

The expression 'value free sociology' was created by Weber (Weber, 1949) in an attempt to establish a less naïve and more sophisticated methodology.

It was taken by him to mean something equivalent to the more positivist notions of objectivity in the natural sciences. (Benton and Craib, 2001. p81)

Weber agrees with the views of the positivists that a fact-value distinction ought to be preserved, and social science should only concern itself with questions of hard-data or facts, while remaining ethically neutral on questions of values. Weber argues that an adequate description of social practice requires us to understand the meaning of the practices to the agents involved, which, in turn, also presupposes an understanding of values (which demands the implementation of the 'verstehen' sociology). Thus, Weber insists that the researcher must understand the values of agents and consider both the subjective and objective dimensions of social life.

Rosaldo (1989), writing in the context of the Weberian tradition, criticises the identification of detachment with scientific objectivity and the myth of the observer as a 'tabula rasa'. He argues that it is rare, if not impossible, for a researcher to become truly detached. Rosaldo argues that Weber's advocated neutrality does not exclude the scientist's passion and enthusiasm, and that the Weberian perspective underestimates the analytical capability of feelings of anger, frustration, depression, passion etc. and results in the elimination of other valid sources of knowledge. Rosaldo concludes that the researcher is a *'positioned subject'*, whose *'life experiences both enable and inhibit particular kinds of insight'* (Rosaldo, p. 19). This argument implicitly acknowledges the role of the researcher in the research process, and reintroduces the self in social research.

The main coupterpoint to the *'tabula rasa'* concept of value-free research came from what was termed phenomenology which emerged in the 1960's and 1970's. The prime mover in phenomenological philosophy was Edmund Husserl (Benton and Craib, 2001) who developed the concept of a 'phenomenological reduction' which is an attempt to set aside what we already know about something and describe, instead, how we come to know it. Also described as a way of tracing the process through which we give meaning to the world. 'It involves a suspension of our everyday, common-sense beliefs and an attempt to describe how we come by those beliefs.' (Benton and Craib, 2001. p83)

Subsequent research challenged the idea of a social science in which the experience and values of the researcher had to be obliterated. Opinions differ whether an acknowledgement of the self is all that is required, or whether the self may legitimately be relied upon in the research process. Harris (2001) argues that the self ought not only to be disclosed, but may legitimately be utilised as a source of knowledge. The author discusses how her own life has been affected by non-profit and voluntary organizations, which later led to her involvement in charitable work, and stimulated her academic interest in the subject. Thus the academic work is influenced by the researcher's family history, and the researcher's knowledge is enriched by her life experience and charitable work.

Substituting the self in the research process does not mean that the research should be less rigorous. Ultimately it is incumbent upon the researcher to keep the subjectivity under control and balanced and present and analyse the evidence objectively. With participant observation or insider research, as Jessop and Penny (1999, pp. 213-16) argue, we are always offering a view from "somewhere" and whilst this may not be an "all seeing eye, a view from everywhere" equally we cannot provide a view from "nowhere."

The selection of an embedded case study design (in GSK) poses a particular challenge arising out of the researcher's close involvement with the organisation. This challenge is more pronounced when framed against the backdrop of the 'value-free' research concept outlined above. The author of the study has held the position of Director of Innovation Excellence for the commercial group in GSK. That he has never worked directly in the R&D (division of the) organisation means that the connection with the interviewees in this study has not been in a direct reporting line. Louis and Bartunek (1992) note that understanding and insights about a company can be acquired either by studying data about the company from the outside or by becoming part of the organisation and studying it from the inside.

Since organisations can be viewed as, or are often compared to, societies with their own peculiar customs and practices, participant observation has become increasingly popular in organisational research (Iacono et al, 2007). Evered and Louis (2001) identify two different paradigms of organisational research, and term the two approaches 'inquiry from the outside' and 'inquiry from the inside'.

The former demands that the researcher be separate and apart from the organisation; while the latter is characterised by some level of personal involvement by the researcher in the context of the research.

Familiarity with a firm can be acquired in two ways: by examining data published by the organisation e.g. company files, financials etc. (i.e. - enquiry from the outside) or by the researcher's presence within the organisation (enquiry from the inside) and *being there*, becoming immersed in, and part of the phenomenon being investigated. The authors (Evered and Louis, 2001) reflect upon their own personal experience entering an unfamiliar organizational setting. They describe becoming aware that, despite their training in the scientific method, they were adopting a different mode of enquiry to make sense of the new organisation: *'It was a multisensory, holistic immersion'* (p. 387) whereby the authors were *'noticing acutely'*.

They report that they did not test hypotheses, but relied on improvisation learned in practice. Published academic research offered little guidance in understanding the new organizational setting, whereas papers by industry practitioners appeared more meaningful and relevant. The authors conclude that the knowledge acquired through *'inquiry from the inside'* is inherently more valid and relevant to the organisational actors. Management research presents challenges of its own. Managers are busy individuals, and are typically reluctant to allow access unless they can see some benefit to the organisation. Hence, access for fieldwork may be difficult to obtain, and, if granted, it, very often leads to helpful insights about complex practices.

A major criticism often levelled at participant observation is the potential lack of objectivity, as it can be argued that the researcher is not an independent observer, but a participant in the phenomenon being observed. The notion of participant observer does presuppose a degree of emotional detachment from the subject matter, the clear objective of the researcher being the conduct of the research.

Inevitably participant observation raises ethical dilemmas: the investigation should not be conducted in a covert manner; informants should be informed of the nature and scope of the investigation. On the other hand, participant observation carries with it the concern that the presence of the investigator may influence the way informants behave. Informants may be suspicious of the researcher and reluctant to participate or be eager to please; they may interject their own impressions and biases etc. The personal relationship between researcher and informants may also influence the interaction (e.g. the researcher may empathise with his/her informants and vice versa). This ought to be taken into consideration when conducting the fieldwork. It is incumbent upon the researcher to build a relationship based on trust, and collect, analyse and display the evidence objectively.

Coghlan and Brannick (2005, p. 61) support insider-research saying that it facilitates' the knowledge, insight and understanding of organisational dynamics, but also the lived experience of one's own organisation.' All of which are difficult to replicate with much legitimacy as an outsider. In the current research, it was a particularly important consideration and challenge to maintain objectivity and neutrality.

In this study the self, or author as participant, is not only explicitly acknowledged (for example, in this chapter discussion), but this dual role of participant and researcher is also used as a source of knowledge. The author's professional experience influences the research process, from the initial choice of the research topic and of the research method, to the presentation and interpretation of the findings. The academic work is enriched by the experience of the practitioner, and the reliability of the findings can be argued to be enhanced by the credibility of the researcher as an industry insider. The academic work in turn gives the author, through the collection and analysis of empirical data, further insights into the nature of innovation in large, global, R&D-intensive firms. During the course of the research project the author becomes more critical and reflective, increasingly aware of his reflecting in action, and better able to reflect upon this reflection in action and thus better able to articulate the tacit knowledge derived from the research.

In the research project, strong and deliberate efforts were made to minimise concerns over subjectivity. This was partly accomplished by relying on multiple informants; by including interviewees from different levels within the organisation as well as integrating extensive evidential support from documents and archives in order to support the findings which will be presented in the next chapter. Other elements included incorporating vignettes of practice, videos, close-out reports, quotes from company files etc., and letting the facts speak for themselves; analysing the evidence objectively through within-case and cross-case analysis, comparison with the extant literature, triangulation of data sources and of theories and, finally, alternating between inside and outside enquiry, especially Chapter 4 which provides an in-depth analysis of both the industry and the company itself.

3.7   Summary

In this study a strong emphasis was placed on developing an effective research design. This was necessary, because many management factors influence an organisation's capacity to produce radical innovation. In consideration of the research objectives it was concluded the research strategy had to be qualitative, using semi-structured interviews, which allow for the collection of rich deep data and the emergence of a relationship between theory and the findings. This facilitated an interpretive, inductive study, a mode of investigation largely missing from the field of disruptive innovation. Moreover, as innovation is often mediated through teams, it is important that the research design is capable of finessing subtle but significant interaction which may either encourage or inhibit innovation. Key elements of the research design can be summarised with the following four points:

1) As the process of innovation cannot be cleanly extracted from its context (organisational, team, individual or project), a decision to use qualitative, inductive research seems warranted.

2) The ISF project offers a rich opportunity to study the innovation process within a bounded and unique organisational experiment which makes it a revelatory case study.

3) The longitudinal aspect of the case allows for an in depth evaluation of precisely which processes had impact on the innovation outcomes.

4) That the case study data explores three layers of management seniority within GSK (from C-level down) adds to the richness of the resulting insights.

This case study offers a rare and privileged lens into a large, complex, global, R&D-intensive organisation. The specific design of the ISF initiative itself, lends itself particularly well to case study analysis as it has been designed as an internal tournament for radical innovation ideas and it is time bound with the company's own pre-defined success criteria built in. Such a project design allows types of analysis which would not otherwise be possible.

This inductive study seeks to develop insights from managerial practice through which to answer the research questions about how innovation happens within complex organisations. The study is qualitative for a number of reasons; the participants in the project represent a group sufficiently small to allow almost a census of members and it is felt that the qualitative approach would allow a deeper, richer interpretation of the phenomena being studied. In depth interviews were favoured as the approach most likely to yield authentic and practical insights from the individuals involved. The study is a twin case study as it follows two separate teams which were competing against one another to develop ideas for radical new products and services for GSK.

The case study also chronicles three separate but connected sequences of events: 1) the set of circumstances which led to the setting up of the project, 2) the experiences and reflections of the team members who participated in it and 3) the views and evaluations of the judging panel who were charged with assessing the quality and usefulness of the ideas produced. The selection of a case study analysis also allows for the inclusion of various types and sources of data. Many secondary sources were available to the researcher to enrich the primary data with important context.

Chapter Four
Background to the Pharmaceutical Industry and GSK

4.1   Introduction and Purpose of the Chapter

Studies suggest that teams are inevitably a product of the context in which they operate (Mathieu et al., 2008; Sarin and O'Connor, 2009). The ISF team, the subject of this case study, was working within the R&D division of one of the world's largest pharmaceutical companies, GSK. Hence an overview, not just of the company itself but of the industry in which it is competing is warranted.

The ISF project was initiated in 2007 and hence this analysis focuses on that year in particular.

This chapter begins with a brief overview of the competitive landscape and dynamics of the pharmaceutical industry at the time of this organisational initiative. It explains how the industry was, at that time, coming to terms with falling success rates and rising costs of R&D. It outlines some of the measures companies undertook to supplement or offset the prospect of declining revenue from their R&D pipeline. Among the measures described is the diversification of many companies' portfolio of products. One such measure involved the re-orientation of R&D away from small molecule medicines targeted at conditions affecting significant portions of the population, towards the discovery of medicines to treat specialist, niche conditions. These treatments are generally more advanced and sophisticated, than conventional medicines, using biologics and vaccines. This strategic approach, which was gaining sway in 2007, moved the pharma business model towards selling fewer products but at a higher price point.

One of the sources quoted in this introductory piece is Dr JP Garnier. Dr Garnier was the chief executive officer (CEO) of GlaxoSmithKline between 2000 (from the original merger between SmithKline Beecham and GlaxoWelcome) for eight years to 2008. As head of the world's second largest pharmaceutical company, he dedicated himself to renewing the company's engine of R&D and he wrote and spoke prolifically about this topic. As he presided over his final AGM in May 2008, the Guardian (newspaper) wrote:

> Few industry observers doubt Garnier's vision and boldness - many of his reforms at GSK have been copied by rivals and the all-important "pipeline" of new drugs, once one of the worst in the industry, is now widely considered to be one of the best. (Guardian, Nick Huber, May 20th, 2008)

Garnier saw the increasing complexity and large scale of R&D operations being the culprit for many of the industry's (and the company's) problems. He attempted to foster a more innovative, entrepreneurial and passionate culture in R&D. An element of his internal programme aimed at making the GSK corporate culture more entrepreneurial cascaded into the Consumer Healthcare division and encouraged initiatives like ISF.

The chapter then moves to a more context-specific description of GSK's business and organisation. The Consumer Healthcare division in which this case study takes place is profiled in detail. In-company interviews are combined with external financial analyst reports to illustrate the central role of new product development to the success of the company; a role which was reflected in the company's strategic priorities. It is in this context that the ISF project emerged as a vehicle to deliver the radical innovation that the company so keenly sought for its pipeline.

The chapter then describes how the company organises for innovation, generally. The ISF initiative was supplementary to the regular, established innovation process within the company and it is this latter process, which is explained here. Aside from the ISF project, the company had an elaborate and successful innovation programme and infrastructure and this is described in detail in this chapter. This profile of the innovation process within the organisation will help readers understand the context in which the ISF initiative was conceived, in the first place, and what made it different to the way innovation is conventionally managed in GSK Consumer Healthcare (CH).

4.2   A Brief Overview of the Pharma Sector in 2007

One simple fact defines the pharmaceutical industry: 'The key to long-term growth has to be R&D pipeline success' (McNamara, 2004; p 25). Garnier (2008) stated:

> The business model of Big Pharma is straightforward. New products are discovered, developed, launched, and protected by various patents. Initially the products benefit from monopolistic – or at least oligopolistic – pricing. After 10 or 12 years, in general, patents expire and lower-priced generics come in, wiping out the revenues of blockbuster drugs in a matter of weeks. R&D must continually replace older products with new ones to stop the revenue base from shrinking. (Garnier, 2008, p. 70).

The pharmaceutical market, in 2007, was going through a period of considerable change. Its traditional reliance on what are called 'small molecule', *blockbuster* (products with annual sales of US$1bn or more) compounds was beginning to wane. Bain and Shortmoor (2010) reflect that market fundamentals had been changing with the erosion of the blockbuster-drug model, traditionally supported by small-molecule drugs (i.e., drugs with a molecular weight of< 500 Da), in favour of an increased emphasis on biologic-based development. The FDA explain as follows:

Biological products include a wide range of products such as vaccines, blood and blood components, allergenics, somatic cells, gene therapy, tissues, and recombinant therapeutic proteins. Biologics can be composed of sugars, proteins, or nucleic acids or complex combinations of these substances, or may be living entities such as cells and tissues. Biologics are isolated from a variety of natural sources - human, animal, or microorganism - and may be produced by biotechnology methods and other cutting-edge technologies. Gene-based and cellular biologics, for example, often are at the forefront of biomedical research, and may be used to treat a variety of medical conditions for which no other treatments are available. (FDA Website, 2009; http://www.fda.gov/AboutFDA/CentersOffices)

The 'Big-Pharma' business model has traditionally relied heavily on small-molecule products which were developed to prevent or treat illnesses and conditions that affected significant portions of the population. Once the development costs were met, these drugs were relatively inexpensive to manufacture and distribute thus allowing companies to maximise revenue through aggressive marketing and selling campaigns. However, once patent protection expires, these medicines are relatively easy for the generic manufacturers to replicate and proprietary manufacturers face what the analysts refer to as the 'patent cliff'; a term which graphically describes the dramatic sales decline for proprietary brand pharmaceutical products once generic competition becomes available. The resulting commoditisation of the small-molecule market has forced the Big Pharma players to seek diversification into areas of high unmet need (e.g., oncology) or, in terms of molecule type, into biologics, vaccines or consumer healthcare. Another way pharma companies found to supplement revenue from a disappointing pipeline was to focus on their patent protection strategies and ensure they secured maximum-marketed portfolio revenues (Table 4.1).

At this time, some companies were rebalancing their portfolios away from the original block-buster model, in favour of vaccines; a market segment which was seen to have higher growth potential. Consumer Healthcare (a term usually used to describe a portfolio of products or a division within large pharma companies which is concerned with the marketing of healthcare products which can be sold without prescription or over-the-counter OTC) was simultaneously becoming a high priority for some pharma companies. Paradoxically, in other companies, the reverse was happening and they considered Consumer Healthcare a distraction to their core business. Hence, some companies were selling their consumer healthcare divisions but even more companies were queuing up to buy. Three large consumer healthcare companies came on the market between 2006 – 2007 (Roche Consumer Healthcare, Boots Healthcare and Pfizer Consumer Healthcare) and, in each case the sales were over subscribed and had to go to sealed bid auction to determine who would secure the purchase. It can be inferred from this that although some companies were divesting, even more were trying to acquire.

Table 4.1: Factors Inspiring the Shift Away from the Primary-Care Blockbuster Model towards Niche Indications

Shift away from primary care blockbuster model..	...towards niche patient populations
•Long, expensive development model, costing $1bn on average •Competition from me-too drugs •Generic versions of numerous primary care drugs already or soon to be available •Shrinking primary care salesforces •Most drugs only work in approximately 50% of patients •Risk of side effects and pharmacovigilence issues	•Faster and often cheaper research and development processes allowing for a greater time at market •Smaller salesforces required •Orphan drugs receive incentives such as market exclusivity, tax and fee reductions and regulatory assistance •Evolving understanding of the etiology of rare diseases •Greater focus on speciality and personalised medicines targetting smaller patient populations with high value drugs •But competition for reimbursement is set to intensify.

Source: Datamonitor.

In 2004, concern began to be reflected in the industry about the 'R&D crisis in the pharmaceutical industry' (McNamara, 2004. p. 18). Garnier (2008) noted that from December 2000 to February 2008 the top fifteen companies in the pharmaceutical industry lost approximately $850 billion in shareholder value, and the price of their shares fell from thirty-two times earnings, on average, to thirteen. Garnier (2008) argues that although analysts attribute the loss of confidence in the industry's performance to a 'perfect storm' of issues like rising costs, stringent regulatory requirements, high-profile product withdrawals on the grounds of safety concerns; he believes 'that declining R&D productivity is at the center of its malaise.' (Garnier, 2008, p. 70)

PharmaWatch (2008), a part of Datamonitor, in their review of 2007 specifies the conditions which made it such a difficult year for pharma companies. First, they forecast that, between 2007 and 2012, the top fifty pharmaceutical companies will face patent expiries on $115 billion worth of drugs. The report stated:

Furthermore, in 2007 the FDA approved just 19 new products: the lowest level for over twenty years. At the same time, the cost of developing innovative therapies keeps rising, reaching an average of $800m to bring a drug to market: fifteen times higher than that recorded in the 1970s and more than three times higher than in the 1980s. This lack of true innovation, coupled with recently increased regulatory scrutiny and tougher cost-containment measures from payers to drive down prices, has made the healthcare environment more difficult to operate in than ever before. (Datamonitor, 2008; Market Watch Biotechnology; p. 19).

Coinciding with this was growing acknowledgment in the pharmaceutical industry generally that falling R&D productivity was a major factor affecting prospects for future company revenue growth and investor confidence. The major expenditures for pharmaceutical companies occur in R&D and marketing or promotional activities. Within R&D, productivity is measured by calculating the clinical quality, sales potential and number of pipeline products being developed relative to the time and money invested in their development.

According to the Pharmaceutical Research and Manufacturers of America (PhRMA annual report 2002–2003), US pharma companies' global R&D expenditure increased from an estimated US$2bn in 1980 to more than US$30bn in 2001; the equivalent of a year-on-year growth rate in R&D investment of 13.9 per cent for the two decades. Over the same period of time, ethical sales grew by a compound annual growth rate (CAGR) of 10.8 suggesting that the growth in product sales is below that of R&D investment and therefore, the calculation concludes; R&D productivity is declining.

The decline in productivity appeared to be as a result of two factors; higher R&D costs combined with a lower success rate. McNamara (2004) argues that the success rate in the development of new chemical entities was declining. She quotes absolute success rate figures of just 0.1% - 0.2%. With success rates that low from the R&D pipeline, she notes that companies have become heavily reliant on a few compounds with sufficiently high, global sales to recoup development costs and provide resources for future R&D expenditure. This gave rise to the majority of pharmaceutical companies basing their revenue growth model on high earning products such as blockbusters. Companies also sought to extend the patent life of their big-brand, blockbusters by introducing minor line extensions, such as sustained-release, in order to fend off generic competition.

As internal R&D fails to generate the breakthrough ideas needed to create blockbuster medicines, pharma companies were increasingly looking outside their own labs to engage in Open Innovation partnerships, joint ventures, acquisitions and alliances. According to Datamonitor (2009), the key drivers for this R&D licensing trend centre on: addressing the current R&D productivity crisis; overcoming patent expiry issues; matching R&D products with the company's corporate strategy; and accessing enabling R&D technologies, which are too costly to support in-house.

Smith (2009) points out the industry signals which have heralded the pharma R&D crisis for some years have been many and obvious. Slowing sales growth, low cost competition and incremental, rather than revolutionary, product development are all, he argues, characteristic signs of a maturing industry life cycle. From these phenomena, he infers that the current model of the research-based pharmaceutical company is moving from maturity to decline.

Another issue commonly observed to be a barrier to R&D productivity and contributing to this decline, was the size of the R&D operations of the pharma companies. There was an industry view that companies had allowed their R&D teams to get too big and that the bureaucracy inherent in such scale was slowing them down. Again, Garnier (2008) sums up the view:

Another culprit is the enormous size and complexity of the traditional pharmaceutical R&D organisation. In drugs, electronics, software, and other industries where fundamental discovery (as opposed to continuous improvement) is the key to success, size has become an impediment. (Garnier, 2008, p. 72).

Garnier is emphasising here that 'fundamental discovery' is different to incremental innovation and it is in the quest for radical innovation, in his view, that size becomes a barrier to innovation. His approach in GlaxoSmithKline to reducing the bureaucracy and enhancing the innovative culture inherent in large, diversified R&D divisions was to split the R&D group into centres of excellence in drug discovery (CEDD's). He argued:

The way to solve the productivity problem is not to break up the pharmaceutical giants into smaller companies. It is to return power to the scientists by reorganizing R&D into small, highly focused groups headed by people who are leaders in their scientific fields and can guide and inspire their teams to achieve greatness. It is to seek the best science wherever it resides, inside or outside a company. It is to fix broken processes and promote a strong culture of innovation marked by a passion for excellence and awareness that results matter. (Garnier, 2008, p. 73).

Smith (2009) identifies another strategic trend called "repositioning" in the pharmaceutical industry which was being explored at that time. This describes the process where the industry R&D heads have begun commissioning specialist biotech firms to put their previously discarded experimental compounds, some that failed in clinical trials as long as twenty years ago, through a series of new tests. The hope is that a medication intended for one treatment turns out to be efficacious for some other condition. There are a number of instances where this apparent serendipity has worked; Pfizer's Viagra was originally developed as a cardiac treatment; GSK's Zyban was developed as an anti-depressant but found to help people quit smoking; and, Pfizer's Vareniclene was intended as a weight loss drug but it also was found to help people quit smoking. All of them managed to secure sales of $1bn at their peak.

Finally, pharma companies in 2007 were increasingly looking at acquisition as a means of boosting the assets in their own R&D pipelines. The Financial Times, in a review article concludes:

Of the three ways for pharma giants to deploy their ample cash flow – pay it out, invest in research and development or buy someone – the last one makes most sense in the current low-return environment. R&D has not been a strong point for quite some time. (FT, 12th October 2010).

In 2007, there were 11 mergers and acquisitions in the global pharma market but according to analysts, only two were of global scale and significance. Those involved, US-based Schering-Plough and UK-based AstraZeneca.

Consumer Healthcare also became an area of increasing focus for pharma companies in 2007. It was widely thought that these businesses, with their well-known and durable brands, represented a more steady and (relatively) certain income stream in an industry where, in the prescription business, patent expiry could precipitate enormous changes in company value and prospects. In 2006, Bayer bought the Consumer Healthcare division of Roche Pharmaceuticals. Also in 2006, the Consumer Healthcare division of Pfizer was sold to Johnson and Johnson. Many analysts believed that Pfizer disposed of its consumer healthcare division in order to bridge a revenue deficit which arose from the patent expiry of Lipitor (a cardio-vascular drug) which was going to leave a hole of $6bn on Pfizer's balance sheet. Datamonitor's (2009) review of Pfizer, focuses on the drought that afflicted the Pfizer pipeline between 1998 and 2004, during which period no major products were launched; they suggest Pfizer masked this lack of innovation with M&A activity.

Similarly, in 2007, Reckitt-Benkaiser succeeded in acquiring the consumer healthcare portfolio of Boots healthcare. All three of these acquisitions were the subject of high profile, well subscribed auctions. These acquisitions and disposals helped illuminate the various strategies being pursued by the large players. Roche had decided to divest of consumer healthcare products in order to concentrate on biologics and cancer treatments. Bayer, the original makers of Aspirin, had declared their strategy of being a top three player in consumer healthcare and so the sale was a perfect match. Boots (Healthcare) decided to concentrate on its retail operations and was not intending to globalise the brands they had in the consumer healthcare portfolio (brands like Nurofen, Strepsils and Optrex). These brands were very attractive to Reckitts who already had successful brands like Lemsip and Disprin and wanted to increase their scale in this segment.

2007 was a year which saw a growing recognition of the decline of the traditional pharma business model and many observers concluded that it was the deterioration in productivity from R&D which was the main culprit. The major pharma companies embarked upon a series of diverse initiatives in order to offset the inevitable fall in revenue which would result from a decline in R&D productivity.

Some companies re-engineered the R&D programme to focus more on higher-value, more specialist medicines like biologics and vaccines. Some tinkered with incremental innovation and even litigation to defend their molecules against generic competition, for as long as possible. Others tried to compensate for the deficit in their internal R&D pipeline by agreeing joint ventures and licensing arrangements with smaller biotech firms.

Despite many in the industry concluding that big-pharma's scale was no longer an asset but actually a liability in bringing new compounds to market; still some companies tried to grow by significant acquisition. But, generally, the acquisition activity was down on previous years. Certainly, what the industry calls 'Mega-deals' were definitely out of fashion. Instead, pharma companies acquired companies who could deliver either strategically important technology or else access to the growing sales opportunities in emerging markets.

Pharma companies increasingly turned to the stability and long-term profitability of the consumer healthcare segment as an attractive means of diversifying their portfolio. As a result, 2006 and 2007 saw an unprecedented flurry of acquisitions in the consumer healthcare business. But even within consumer healthcare businesses with their attractive high and steady revenues, innovation was still a key priority.

4.3   GlaxoSmithKline – A Summary Profile of the Company

GSK is a research-based, pharmaceutical company with operations in over 100 countries and employing over 100,000 people. The company was formed in 2000 from a merger between SmithKline Beecham and Glaxo Welcome and is headquartered in London, UK. GSK is a major global healthcare group engaged in the creation, discovery, development, manufacture and marketing of pharmaceutical and consumer health-related products. While, GSK has its corporate head office in London, it also has operational headquarters in Philadelphia and Research Triangle Park (North Carolina), USA, and operations in some one hundred and fourteen countries, with products sold in over one hundred and forty countries. The principal R&D facilities are in the UK, the USA, Belgium, Italy, Japan and Spain. GSK's products are currently manufactured in some thirty-eight countries. The major markets for the Group's products are the USA, France, Japan, the UK, Italy, Germany and Spain.

GSK operates principally in two industry segments:

• Pharmaceuticals (prescription pharmaceuticals and vaccines).
• Consumer Healthcare (over-the-counter medicines, oral care and nutritional healthcare).

R&D is a significant investment for GSK, in fact, the company is one of the world's top ten R&D spenders across all categories and industries. According to data compiled by the European Commission (NATURE, Vol. 450, 1st November 2007), the pharmaceuticals and biotechnology sector has overtaken every other industry, including technology hardware and equipment to become the leading R&D investor worldwide. Pfizer is currently the largest R&D single investor in the world, with Johnson and Johnson (J&J) and GlaxoSmithKline also in the global top 10. GSK spends approximately 14% of its turnover on R&D (GSK Investors Report Presentation Q2 2010). This, high level of investment in R&D and consequent focus upon discovery and innovation, is a defining feature of both the industry and of the company and make it particularly suitable for inclusion in a study investigating the NPD process within R&D teams. Ornagh (2008) makes the point that in the R&D intensive, pharma industry 'both margins and costs are largely determined by innovation.'

GSK's 2006 Annual Report describes the company's reliance on innovation as follows:

> Continued development of commercially viable new products is critical to the group's ability to replace lost sales from older products that decline upon expiration of exclusive rights, and to increase overall sales. Developing new products is a costly and uncertain process. A new product candidate can fail at any stage of the process, and one or more late-stage candidates could fail to receive regulatory approval.
> New product candidates may appear promising in development but, after significant investment, fail to reach the market or have only limited commercial success. This, for example, could be as a result of efficacy or safety concerns, inability to obtain necessary regulatory approvals, difficulty or excessive costs to manufacture, erosion of patent terms as a result of a lengthy development period, infringement of patents or other intellectual property rights of others or inability to differentiate the product adequately from those with which it competes. (p44)

The Consumer Healthcare (CH) division of GSK is of immense strategic importance to the company overall. The (London) Times wrote in March 2008:

> Glaxo (the entire company) trades on a price-earning multiple of about 11 times earnings while the consumer healthcare companies seem to be closer to 20 times. There's a lot of hidden value there.

Moreover, despite the fact that Consumer Healthcare develops products primarily for sale over the counter (rather than via prescription), the level and quality of R&D required, along with the regulatory oversight applied by national government agencies is equally high to those that apply to the prescription (also called 'ethical') pharmaceutical business. The same regulatory approval body and the same approval process applies to OTC medicines as to prescription ones. It could also be argued that the same rigour and pressure applies to the level of innovation required to succeed in the very competitive OTC and personal care markets.

In 2007, (GSK) Consumer Healthcare was exhibiting positive sales performance with worldwide sales growth up 14% (Table 4.1) in that year. The CH division has a strong brand portfolio that includes *Lucozade, Sensodyne, Panadol, Horlicks* and *Aquafresh*. 2007 saw the US launch of *Alli*, the first over-the-counter (OTC) weight loss aid approved by the Food and Drug Administration (FDA). Alli has subsequently also been launched in Europe. Alli is a major innovation for GSK and is considered by the company to be a breakthrough in the area of self-medication for weight loss. Alli is an example of an open innovation approach in GSK as the molecule was originally launched by Roche Pharmaceuticals as a brand called Xenical; a prescription product to help with weight loss. GSK bought the technology and rights to the brand in the expectation of being able to market it over-the-counter, in various markets around the world. The process of taking a prescription product whose efficacy, but more importantly safety, has been satisfactorily demonstrated and allowing it to be offered for sale without prescription is known as 'switching'. There are a number of examples of pharmaceutical products which have been successfully switched, having originally required a prescription; Tagamet and Zantac are notable examples. Switching generally happens at the end of their patent protection, when products are facing generic competition and it serves to prolong the revenue stream attaching to a product by continuing to use the brand name and making it available direct to the patient.

Table 4.2: GSK Financial Performance in 2007

Financial trends and ratios

Total results	2007		Growth*	2006		Growth*	2005
	£m	CER%	£%	£m	CER%	£%	£m
Turnover – Pharmaceuticals	19,233	–	(4)	20,078	9	8	18,661
– Consumer Healthcare	3,483	14	11	3,147	6	5	2,999
Total turnover	22,716	2	(2)	23,225	9	7	21,660
Cost of sales	(5,317)	8	6	(5,010)	6	5	(4,764)
Selling, general and administration	(6,954)	–	(4)	(7,257)	–	–	(7,250)
Research and development	(3,327)	(1)	(4)	(3,457)	11	10	(3,136)
Other operating income	475			307			364
Operating profit	7,593	3	(3)	7,808	17	14	6,874
Profit before taxation	7,452	2	(4)	7,799	19	16	6,732

Source: GSK Annual Report 2007.

At £3.5bn (annual sales), GSK Consumer Healthcare is the Number 2 and fastest-growing consumer healthcare business in the world. The figures above show sales up in 2007 by 14% (and again, in 2008, by another 14%).

Through GSK's Consumer Healthcare business, and its expertise in sales and marketing, its global footprint, the company is deemed to be well placed to be the partner of choice for 'switch' products, bringing them from the prescription to the OTC market. As noted above, many pharma companies do not have consumer healthcare divisions and hence when they have a molecule or medicine that is a candidate for switching, they are likely to look for a suitable partner for the venture.

GSK CH has a number of leading brands in its portfolio (Table 4.3)

Table 4.3: Products Turnover - 2007

Product	£m
Lucozade	347
Aquafresh	308
Sensodyne	293
Panadol	262

Horlicks   174

Source: GSK 2007 Annual Report

The 2007 Annual Report for GSK states:

Global healthcare markets are in a state of change. For example, there is an increasing trend for governments to cut state healthcare costs by influencing a switch from prescription to generic or OTC products. Looking ahead, healthcare is becoming more consumer-centred. Research shows people expect to be able to access medical knowledge and to influence their own treatments. For many, OTC products are their first destination for everyday healthcare. The company expects that the highest rates of growth for all healthcare businesses will be driven by the developing, emerging economies. OTC is the foundation of healthcare in these countries. In China, for example, OTC accounts for 36% of drug expenditure, compared to 8% in North America and 10% in Western Europe. (p.5)

This recognition of the role of the trusted (CH) brands, as a first step in primary care in the vital, emerging markets in Asia has appeared to make the consumer healthcare division more attractive to the parent company than it had been before. These are the markets from which the organisation expects to secure most of its growth in the coming years and hence the consumer healthcare brands could possibly provide a useful bridgehead for driving the business.

The Consumer Healthcare and Pharmaceuticals businesses are not autonomous, separate businesses, but are intended to be complementary and synergistic in a number of important areas. They are both, according to the company's 2007 Annual Report, backed by science-endorsed strategies and have a focus on R&D. The growing worldwide trend for patients to manage their own healthcare (mentioned above), first choosing OTC products, before consulting a doctor is a behaviour that signals positive future for the consumer healthcare brands. The company wants to be able to draw on these skills and knowledge in its Pharmaceutical business and share costs and resources. Consumer and Pharma also share expertise and resources in other areas, such as regulatory matters, R&D, marketing, HR, IT, distribution and procurement.

This chapter's introduction discussed the issues and challenges facing the pharmaceutical business in 2007. While consumer healthcare, to some analysts, looked more stable and reliable in terms of revenue potential; this business was not without its own challenges as outlined in the company's 2008 Annual Report (p. 23):

The environment in which the Consumer Healthcare business operates has become ever more challenging: consumers are demanding better quality, better value and improved performance; retailers have consolidated and globalised which has strengthened their negotiation power and cycle times for innovation have reduced.

But despite these challenges, the Consumer Healthcare business with its established and successful brands has offered some stability to the Pharma parent company. The consumer business accounts for 18% of the total company turnover. A UBS Investment summary published in January 2009 (UBS Investment Report on GSK, January, 2009), suggested that GSK Consumer Healthcare is an 'under appreciated business in a robust market.' It is worth pointing out that this UBS report is unique and a first for GSK. Gary Davies, (in an interview with the author) GSK's Director of Investor Relations says:

Traditionally, the pharma analysts for the investment banks review our (GSK's) performance and they generally append some cursory analysis of the Consumer Healthcare business. In the case of this UBS report, the analyst who prepared the report spent time with the senior executives in the company was a dedicated consumer healthcare and FMCG company analyst. Therefore the UBS report is the most detailed investment report ever into the CH business in GSK.

UBS is a 150-year-old Swiss investment management, wealth management and investment analyst firm. In their in-depth report on GSK (published on January 6th, 2009) they note:

Demographic and lifestyle changes, emerging markets opportunities, broadening of distribution channels, increased innovation, and Rx-to-OTC switching will in our view cause an acceleration of industry growth to 5% in the coming years. With GSK one of only two global players (the other being Johnson & Johnson) with high capabilities across our four key value dimensions, we believe GSK can perform well above the industry.

Thus, the analysts believe that the Consumer Healthcare business is a significant and currently undervalued part of GSK. They note:

While the perfect storm (declining R&D productivity, patent expiries, and pricing and profitability problems) rages in pharma, GSK has diversification (through the Consumer Healthcare division) that should partially insulate GSK.

When reviewing the Consumer Healthcare company's performance and future prospects, it is for innovation that the company receives some of its highest marks from UBS (Figure 4.1). UBS evaluate GSK's prior performance and current capability against a number of competitor companies along four key criteria. These four criteria are explained in more detail in Figure 4.2 and are: new business opportunities; innovation; geographical footprint and; marketing excellence.

Figure 4.1: UBS Analysis of GSK Consumer Healthcare

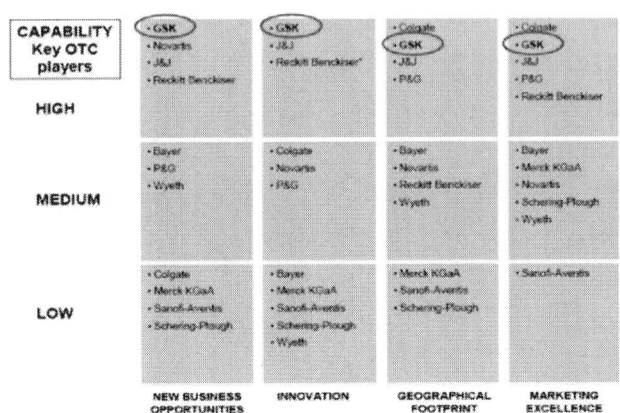

Chart 1: Global OTC players - qualitative assessment of capabilities along key value dimensions (UBS)

Source: UBS Report on Global Consumer Healthcare Market and Competitors.

UBS explain the definitions behind their rankings with the following explanatory chart:

167

Figure 4.2: UBS Explanation of Units of Analysis

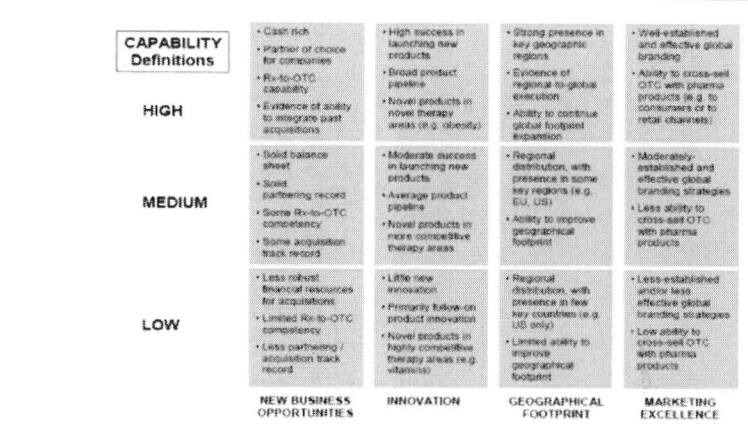

Source: UBS Report on Global Consumer Healthcare Market and Competitors.

GSK Consumer Healthcare also has a successful track record in acquiring smaller companies with strong regional brands and developing those brands globally. The acquisition of Sterling Health in the early '90's propelled the Panadol brand, from a regional analgesic brand into a global brand leader in pain relief. Similarly, when GSK bought Block Drug in 1998, it was able to launch the Sensodyne brand for dentine hypersensitivity in many new markets, most recently in China (2008). Similarly, in 2007, GSK bought CNS, a local US manufacturer of *Breathe Right* nasal strips and within the first year of ownership, the brand had been launched in 46 new countries. The company prides itself on being able to rapidly integrate and maximise the value of acquisitions.

4.4   The Consumer Healthcare Business

The GSK Consumer Healthcare business operates in three main product or therapy areas:

- Oralcare – i.e. toothpaste, toothbrushes and associated products.

- OTC – Over The Counter medicines are those used to treat mild, self-limiting ailments which can be purchased without a prescription.

- Nutritional Healthcare – Nutritional drinks and health supplements.

When separated out, the consumer healthcare sales performance, including 2008, is shown as follows in table 4.4.

Table 4.4: Composition of GSK Consumer Healthcare Annual Sales 2007

	2008 £m	2007 £m	2006 £m
OTC Medicines	1,935	1,718	1,496
Oral care	1,240	1,049	993
Nutritional Healthcare	796	716	658
Total Consumer Healthcare	3,971	3,483	3,147

The cornerstone of the GSK CH's strategy is innovation. The company's president, in 2008, published a model outlining the strategy which was circulated to all senior managers (see Figure 4.3).

Figure 4.3: GSK's Strategy 'Cathedral'

Source: John Clarke's Briefing Documents for Senior Managers in GSK, 2008 (GSK Consumer Healthcare Internal Strategy Document, published by the Divisional President, 2008).

As can be seen from this model, the very first pillar of the strategy relates to improving the company's ability to innovate; the primary growth strategy is to 'Accelerate Innovation'. John Clarke became President of the Consumer Healthcare business in 2006; he is, and has been, completely unequivocal in his commitment to innovation. In the Annual Report of his first year in the position he said:

> We expect to achieve this growth through a vigorous focus on delivering new product developments that are tightly aligned with consumer needs.

In an interview with the (London) Times on 28th March, 2008, he reiterated:

> Consumers have a high interest in innovation. It's a consumer driven aspect of our market. The moment you become an inferior product, you are in trouble. Loyalty does persist and may carry your brand for a few years, but you must innovate to have better products.

GSK is the number three Oralcare company in the world (Figure 4.4) behind Colgate and P&G (who market the Crest brand). It is also the number three player in the OTC market behind J&J and Bayer. In Nutritional Healthcare, the company is the market leader.

Figure 4.4: Euromonitor Global Rankings in OTC and Oralcare

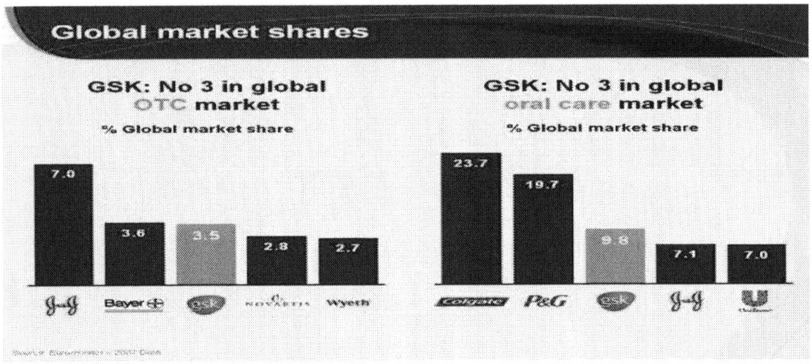

Source: Euromonitor 2007.

4.5 Structuring the Company for Innovation – The Launch of the Future Group in 2004

In 2004, the company announced a new business and organisational model designed to deliver better top-line growth with the objective of enhanced focus and simplicity. The company's new vision underpinning the changes was: 'To be the fastest-growing consumer healthcare company, through innovation, centred on consumers and delivered by science.' The principal rationale and drivers for the change were explained to be as follows:

'Innovation Delivered – We will be the best at creating compelling, differentiated ideas for our brands that we will develop and launch with the urgency of entrepreneurs.

Compelling Communications – We will be the best at connecting with consumers and healthcare professionals through insightful and relevant communication.

New Business – We will be the industry's leader in generating new business from in licensing, acquisitions, new categories and switch (i.e. switching from prescription status to OTC).

Point of Purchase – We will maximise the availability and visibility of our brands for the shopper at purchase decision points.

World Class Operators – We will execute better, faster and more efficiently than our fiercest competitors, while reducing non value-added costs.'
(Shaping Our Future; GSK Internal Publication, 2004)

The most significant development signalled by this new structure in the business was the creation of a series of 'Future Groups', which were physically located in the business' two HQ's in the UK and US, and have a remit to manage the company's biggest brands, globally. These teams have a mandate to 'deliver more innovation faster'. The brands for which a Future Group was established were those whose 'sales and market share were global and whose positioning was sufficiently similar such that they could be most efficiently developed using a global team'. These brands represented 40% of the company's turnover.

At the time of the ISF project, three years later, GSK Consumer Healthcare still operated this model for centralised marketing and R&D unit for all of its brands that are marketed globally and have annual sales in excess of $250m. Seven of the company's brands qualify under these criteria. This unit was still called the Future Group and it retained central control of the overall brand and marketing strategy, new product development pipeline and communications for the company's most profitable and most valuable seven brands. Each of the seven brands has a dedicated team. The responsibility of the team is global. Hence, each Future Team is responsible for creating a global strategy for the brand and, further, for developing new products under the brand franchise with the capacity to be launched globally. On the commercial side, the team comprises people who are experts in insight or consumer understanding, ideation (generating new ideas, refining , prototyping and researching them), innovation, advertising, digital marketing, marketing to professionals (doctors, dentists etc) and brand strategy. Some of these positions are mirrored in the R&D side with experts in new product research and others who are experienced in and adept at product development. Decisions about the strategy, the NPD pipeline, the advertising for these brands are all taken centrally and then implemented locally by the individual markets or country marketing teams. Future Group brands are sold in 120 markets (or countries) around the world. The Future Teams are located in newly created 'innovation hubs' where new workspaces have been created with the intention of maximising the interaction between colleagues. Each brand team has a 'hub'; a space without physical offices or partitions and where everyone is expected to move around and work in a different space each day. Two of these centres have been created, one in the New Jersey (Parsippany) and the other in Weybridge, Surrey.

The 2006 GSK Annual Report alludes to a new structure within the Consumer Healthcare R&D teams 'whereby for the Global brands, R&D mirrors the commercial structure, with brand-dedicated R&D teams paired with commercial brand teams and both located together at the Innovation Centres in Weybridge, UK or Parsippany, USA.' Combining the structures of commercial and R&D enables companies to combine these resources with a view to developing new and enhanced capabilities (Simon et al, 2007).

Leenders and Wierenga, (2008) note that a number of terms have been used in previous research to describe a team of people working together on tasks connected with new product development, for example 'integration', 'collaboration', 'cooperation' and 'harmony' with the common denominator being the joint behaviour towards the achievement of a shared goal. (The term preferred in GSK is 'fusion' which is intended to describe a deep and purposeful collaboration among the R&D and commercial teams.)

In terms of organisation structure, below is how the Future teams fit into the overall GSK hierarchy. The worldwide CEO presides over the Operating Board of Directors, one of whom represents the Consumer Healthcare Division. This President of the Consumer Healthcare Division has a number of direct reports of whom one is the President of the Future Group. Reporting to the President of the Future Group are the VP's of each of the individual brand Future Groups and each of them have a team as noted above.

The ISF project took place under the auspices of the R&D division of GSK. There were roughly 700 people working in R&D in that division at the time. Heading up the R&D function was Dr Ken James and his key direct report for this project was Stan Lech who was head of his group of direct reports called the 'SLT', Senior Leadership Team.

Figure 4.5: Organisation Chart for GSK in 2007

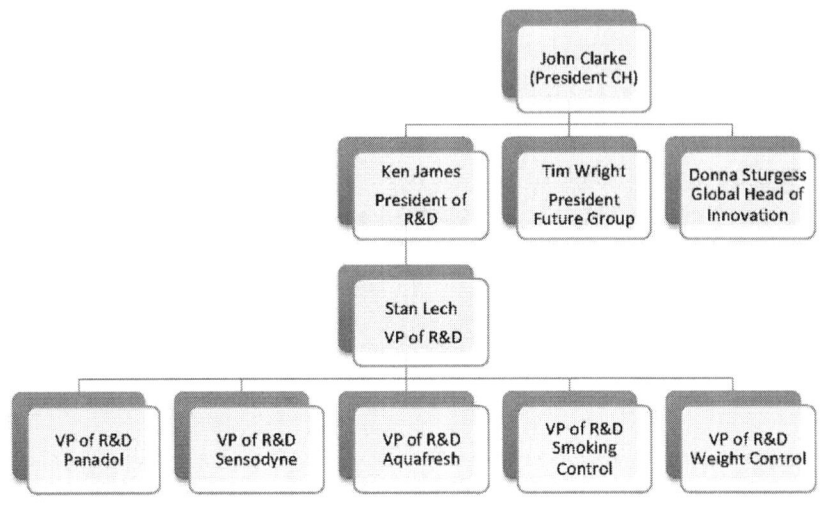

Source: GSK Internal HR Documents 2007

4.6 How Innovation Happens in GSK Consumer Healthcare

In GSK, for the key brands, the Future Group manages innovation. The Future Group manages the process through a phased approach, using cross-functional teams. The figure (4.6) below is one that I have developed (in my role of managing innovation excellence) in the Future Group which is used to brief new employees to the company about how new ideas are managed in the organisation:

Figure 4.6: The Innovation Process in GSK CH

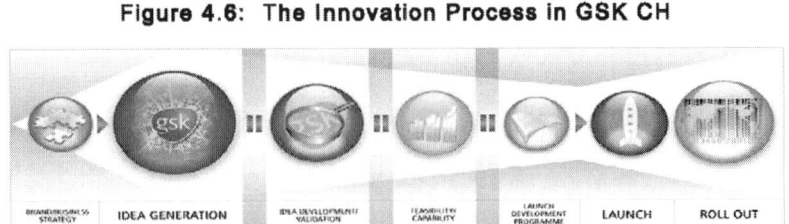

Source: GSK Innovation Excellence Programme 2007 (internal document).

The first phase is designed to ensure that whatever innovation results from the process is 'on strategy' for the brand. This requires a deep immersion in and understanding of the brand, its competitive position, how consumers (buyers) relate to it, what they use it for, how they see it, the dynamics of the markets in which it operates and its history. This phase involves detailed market segmentation for each of the areas in which the company operates where current consumers are mapped as well as the users of competing products and brands. However, the primary focus is to look to areas currently either poorly served or not served at all by the products currently on the market, to identify potential future opportunities.

Then, as the diagram above illustrates, the process becomes a little more divergent as it enters the idea generation phase. This is where, through various means such as:

- Company Suggestion Scheme
- Outside companies or partner firms/universities (now referred to as 'Open Innovation')
- New technologies
- R&D Breakthroughs
- Consumer research
- Brainstorming
- Innovation meetings
- Ethnography
- Lead User Focus Groups
- Online communities

New ideas are received and worked upon by a specialist, Ideation Director whose sole job is to investigate and scope out innovation opportunities and shepherd appropriate ones into the brand's NPD pipeline.

The next phase is Idea Development and it is concerned with fleshing out those new, often not yet fully formed ideas. The Ideation Director has to work with these ideas with the objective of amplifying the elements of them that have most appeal for consumers in that market. Qualitative research is usually used here to establish whether ideas have a genuine heartbeat or not. Simultaneously, the R&D Future Group team will be doing whatever proof-of-principle studies are necessary to ensure that the new ideas are technically feasible in the first place. There is a close working partnership between the NPR Director and the Ideation Director as they work through this stage. The outcome at the end of this phase will be a portfolio of fully finished 'concepts' which are ready for volumetric testing which is a form of quantitative research that explores the likelihood of potential consumers actually buying the product assuming it is to become available. This phase is quantitative and provides the organisation with some certainty relating to the value of the projects being managed within its pipeline.

The next phase is Feasibility/Capability and this is a mainly mechanical phase where evaluations are made about the potential of the project. The commercial attractiveness of most new projects will come down to an evaluation of how attractive the opportunity is when balanced against how difficult it might be to exploit. The company has developed various filters and matrices for evaluation projects and ranking them against each other so that they can prioritise resources.

If a project emerges from this phase, this means that it will be commercialised and the company has confirmed a willingness to make the necessary investment in manufacturing, clinical trials, packaging and the suite of activities required to bring the idea to market. Once a project has reached this stage it is no longer 'fuzzy front end' but is now an established project within the company with its own critical path, project management team and milestones. It moves to another team at this point (although there is overlap for the handover and beyond) and the new team have expertise in global launch programmes. It is their job to maximise the opportunity by planning and executing a global launch programme for the innovation.

The above process is carefully planned, highly structured and reflects the philosophy of having stage gates coordinating the progress of individual projects. The Future Group is the operator of this process for the company. Under normal circumstances, the projects would be initiated by the Ideation Director along with their R&D partner (the NPR Director) and they would steer ideas into the NPD funnel. In fact, this pairing will normally shepherd potential opportunities and ideas right up to what appears on the diagram as the third checkpoint or stagegate. This stage is a meeting with the senior decision makers in the company, in a group called the Portfolio Management Board, where decisions are taken whether or not to progress with further (sometimes substantial) investment in the project.

4.7   The Innovation Sans Frontiers Project (ISF)

Each of the seven major brands (those with annual sales of/approaching $250m) has a dedicated team managing the innovation for that brand. Nevertheless, there was a feeling in R&D that there were some very talented people (principally R&D scientists) who might, with the right stimulus and conditions, deliver more ideas of value to the company. It was also noted that the company's innovation pipeline, at that time, had a large number of incremental opportunities coming through but not enough radical product ideas of sufficient scale, generally measured in sales potential.

There was a proposal to set up a supplementary innovation effort using the most innovative people exclusively within the R&D group. The R&D VP who leads all the R&D effort across all the Future Group brands in the Consumer Healthcare business had been consistently impressed by the calibre of some of the ideas of his people, he cites an example of an individual in his team:

> Take X for example, always known for his way-out ideas that five, ten years on I look back and say, boy! They were very visionary ideas, because that's what we're doing now! But when you first talk to X, you go, Oh! Man, I can't imagine *that* in our business. But also he's been able to rein in some of those ideas and deliver some very concrete projects. So these people flip freely back and forth from creative to delivery, creative to delivery.

One of the R&D VP's who helped select the team members for the project described how the project first came about. The company does an annual talent review for its executives where the senior managers devote a day to discussing the career plans and opportunities for the people under their management:

> There were a couple of things going on at the time. The first time we had a talent review process within the R&D community and we created a category of people that we felt were innovative. And, the question that we discussed at talent review, well if you've got a group of people that identify as innovators. Why don't we utilise their skills in a different manner from classical projects? And, so was really the idea that sort of stimulated the ISF.

To have been identified as an 'innovator' within this group meant that the individuals concerned needed to have shown a capacity to develop creative ideas and have demonstrated a high level of interest in new ideas:

> These people are highly creative, highly energetic, have a lot to bring to the organisation. Having selected them why don't we ask them to work in two teams on identifying new ideas for the company through whatever creative processes they want to? To basically utilise with a completely open brief unconstrained by interference by senior management and to for a period of time work on that and report back with their ideas. But it was really around being creative, having demonstrated that, having identified new technologies, showing an interest in new ideas all of those were consideration factors for the selection process.

There was also a strong feeling at the time that the company was too bureaucratic and that this focus on process was frustrating the creativity of some of the R&D scientists. The leaders of the Future Group R&D teams felt that it would be liberating for the teams to be unconstrained by the traditional regular reporting requirements of the company. Incidentally, the issue of bureaucracy within GSK was one of the first issues raised by the new worldwide CEO, Andrew Witty who took the helm in May 2008 and very soon afterwards (August 14th, 2008) gave an interview to the Economist where he described GSK as a Police State.

It is a rare company boss, let alone one who has just got the top job, that can get away with likening his firm's culture to a police state. But Andrew Witty, the new boss of GlaxoSmithKline (GSK), a British pharmaceuticals giant, somehow manages to pull it off. He invokes that analogy— tentatively, to be fair—to explain the cultural transformation he wants to see at GSK: away from today's excessively regimented, rule-based approach towards the "utopia" of a simplified, values-based culture that trusts employees to do the right thing. (The Economist, 14th August, 2008).

It was against this background, specifically, the following considerations – that the project was conceived:

1. A feeling that there were some creative scientists who, if placed in the right team and with the right conditions, might deliver more ideas of value to the business.

2. The R&D organisation wanted to make a statement to underline the culture of innovation they were determined to promote within R&D.

3. A sense that if the scientists were freed from conventional red tape they might think more creatively.

4. It was believed that this type of initiative would provide valuable insights about the capabilities of the participants that could inform future project team selection as well as promotion decisions.

The creation of this project represented a new and experimental departure for GSK in the area of innovation. It also represented a supplementary effort to the significant and formal work streams that were already an essential part of the GSK structure. Its sponsor in the organisation was the worldwide vice-president of R&D within the company. According to GSK's head of Organisational Development, the management team that inaugurated the ISF team did so because of their shared hypothesis that:

a.　There is value in having competing innovation teams with an open remit and no formal, constraining reporting requirements.

b.　Innovation should thrive under conditions in which a team didn't have to observe any conventional and formal company policies and rules. Specifically, the team did not have to present regular updates to senior management.

c.　The teams were encouraged to work with partners outside their traditional partners for ideas, design and research and this would bring new energy and original ideas into the business and expose the team to fresh thinking.

d.　The teams were hand picked as key talent within the R&D organisation who had displayed strong qualities of entrepreneurship and innovation in previous roles.

The R&D Vice Presidents in GSK Consumer Healthcare (all direct reports of the worldwide head) each nominated a number of candidates to join the ISF teams. The company has two main centres for R&D (as noted above); one in New Jersey, USA and the other in Weybridge, Surrey, UK. They are roughly equivalent in size and it was decided to have two teams, a UK-based one and a

US-based one. Approximately twelve people were selected to join each team. The teams would have an equal spread of talent available to them and would be provided with equal resources including an initial discretionary budget of $250,000 for each team.

Chapter Five

Fieldwork

5.1   Introduction

As described in Chapter Two, in reference to the Innovation Value Chain (Hansen and Birkinshaw, 2007), the innovation process is often described as a journey in which each project goes through a series comprising at least three phases. The phases begin with the process of generating the ideas, either internally within teams or using external experts and other sources, and once the initial raw ideas are captured, the next phase is developing them into distinctive and compelling concepts which are capable of being communicated, both within and outside the company and tested with potential customers. This phase also includes the sifting, ranking and prioritising ideas so that decisions are made about which ideas to progress and which to abandon. At this second phase, teams are concerned with finding the essence of the ideas and amplifying the elements of each idea that might have most resonance with potential customers. They are also concerned with prioritising ideas and are anxious that either they will progress certain ideas that ultimately turn out to be 'lemons' or else abandon others which might have been more promising.

The third and final stage concerns the implementation of the ideas. In process innovation , this might mean the introduction of a better form of production or distribution. In product and service innovation, this phase generally refers to launching or introducing the idea into some form of widespread use or making it commercially available.

The ISF initiative followed the path described in the IVC and hence it seems logical to chronicle the innovation journey in narrative form using the inputs, insights and real quotes of the people involved. Hence the structure of this chapter follows the project teams as they journey through the stages of the IVC within the project. This case study is described in narrative form. All of the sources quoted have been interviewed for this study. The Innovation Sans Frontiers (ISF) programme took pace over a nine-month period and all the participants (except those who left GSK during the prgramme or very shortly afterwards) have provided an in-depth interview for this study. This chapter begins with a desciption of why and how this singular and unprecedented initiative came about. First it follows the story of the innovation journey of the UK team and then it switches focus to the story of the US team. Finally, as happened within the project itself, the two stories merge for the final presentation of the teams' innovation outputs, and the chapter will conclude with the comments of the overall head of R&D within the Consumer Healthcare (CH) division.

The objective of this chapter is to provide an in-depth report of the phases of an innovation project, written in the words of the managers involved. The transcripts of managers' reports of their experiences in the project provide an authentic account of how innovation actually takes place inside oganisations and within teams.

5.2   How and Why GSK came up with the ISF Initiative

5.2.1   October 2006

Ken James, GSK's President of R&D for the Consumer Healthcare division, had called his team together for a meeting in New Jersey. It was October 2006; James had been appointed to this role only two months before and this was a significant meeting for him. The GSK Consumer Healthcare organisation had embarked on a major campaign aimed at improving its performance in new product development (NPD). As part of this, the company had undertaken a significant internal reorganisation in which up to 10% of the global headcount had been lost; the part of the company which is charged with new product and service development was now called the Future Group. So called, because they were charged with 'creating the future' for the company's key consumer brands. 'Creating the Future' was, in this case, essentially concerned with building a robust and well researched innovation NPD pipeline for the company's major brand assets.

The Future Group was divided into two strands: R&D and Commercial. The commercial team were charged with coming up with market-led, customer-driven ideas for new products but it would be the R&D team that would have to bring them to life and ensure that they were technically feasible and capable of manufacture. Consumer insight was the company's conventional approach to generating ideas for NPD but the R&D team were also mandated to actively seek new technologies which might offer GSK commercial opportunities. The R&D organisation employed roughly 700 people at this time.

The commercial organisation, a separate unit to the R&D organisation, had inaugurated considerable change in their ways of working to facilitate the specific future-focus of their role. A new programme to embed a culture of innovation was being introduced into this part of the company. Underlining the cultural change; roles were changed, titles, responsibilities, metrics – all reflected the a higher emphasis on innovation within the corporate culture. This reorganisation within CH was a major effort to make innovation systemic in the organisation and it owed its genesis to the strategy developed by the company's CEO, Dr JP Garnier (described in Chapter Four).

New role titles were aligned with the company's approach to innovation and job responsibilities were more clearly connected with various stages of the innovation funnel or process within GSK. Thus, new commercial roles were developed with 'ideation' in the title to reflect the expertise required in the generation of high-potential raw ideas. This role was involved with research; often referred to as 'customer insight'. It's objective was to use global research to finesse insights which might lead to the generation of promising new ideas with consumer appeal. Th Ideation role required an ability to identify, crystallise and shepherd in new opportunities to the brand and company pipeline.

This commercial 'ideation' role was aligned with the idea generation (mainly) and conversion elements of the innovation process (described in IVC) and this role was mirrored in the R&D organisation by new product research (NPR) specialists and teams. Similarly, as ideas progressed within the organisation, the closer they got towards possible launch onto the market, the more they would be managed through the final stages of the process by conventional project managers. Again, these commercial project managers and teams who were managing these projects towards launch also had counterparts in the R&D organisation who were called new product development (NPD) scientists and teams. Figure 5.1, illustrates both the innovation process (central diagram) employed in GSK as well as the key personnel involved in each phase with the commercial teams shown in above the diagram and R&D illustrated below.

Figure 5.1: The GSK Innovation Process/Funnel

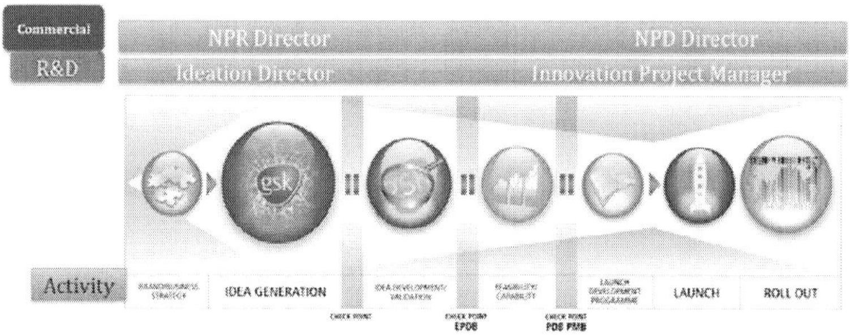

Source: GSK Innovation Excellence Programme 2007 (internal document).

2007 was a period of significant change in the GSK Consumer Healthcare organisation and James wanted to send a signal that the R&D team were keen not merely to play their part but they were enthusiastic about going further to deliver on the organisation's ambition for innovation. He too was engaged in an exercise to rebrand the R&D group as 'Radical R&D' and he was looking for a flagship project that might underline his team's committment to the innovation agenda.

James also had a nagging suspicion that company bureaucracy was stifling the creativity of some of his key people. Now that he was at the helm, he wanted to take the 'brakes of bureaucracy off.' He felt that some talented people were being hampered in making their full contribution in terms of ideas because of the complexity of the company's operating procedures and the high level of managememt oversight. James was curious to see just how much certain creative indivduals might achieve if they were let loose to follow their own ideas.

A third consideration was also preying on his mind, the company's pipeline of new products needed some additional 'excitement'. New product ideas currently being assessed and progressed in the company were adequate but James would like to have seen a few more 'remarkable' or radical ideas. He had the view that the ideas curently in the pipeline were too conservative, too safe and not radical or 'disruptive' and definitely not revoutionary. He knew that the portfolio would always need to reflect a balance between (radical, disruptive) 'innovation' and (safe, incremental) 'renovation' but he felt that the company was veering far too close to the latter category.

That the pipeline was relatively threadbare on the company's most significant brands was the single biggest factor in the decision to create the Future Group, as the commercial organisation, and the Radical R&D organisation to compliment it. The table below lists the senior leadership team supporting the R&D organisation within GSK CH.

Table 5.1: Ken James Senior Leadership Team in R&D October 2006

Name	Role
Dr Ken James	SVP R&D
Bob Wolf	SVP HR
Stan Lech	VP R&D Future Group
Dr Ken Strahs	Head of R&D for Smoking Control Brands
Dr Geoff Clarke	Head of R&D for Panadol
Dr Alexis Roberts Mackintosh	Head of R&D for Aquafresh
Dr Teresa Layer	Head of R&D for Sensodyne
Brendan Marken	Head of New Product Development
Simon Gunson	Head of New Product Research
Jo Moore	Head of Open Innovation
Dr Sandy Lionetti	Director of Organisational Development

5.3   Proposing the ISF Programme

Surrounded by his direct reports, the various R&D VP's, the company's top tier scientists, HR and R&D managers, James explained an idea he'd had which he called *Innovation Sans Frontiers*. The essence of the idea was that he proposed to create a new project within which there would be two R&D innovation teams, one in the UK and one in the US. Each team would be comprised of 10 to 12 existing R&D staff. Each team would have approximately the same level of experience and ability within its membership. That both teams 'would be created equal' was an important tenet of this proposed initiative. What would make these teams different was that they would have all conventional reporting constraints and procedures removed for the duration of the project. In other words, they would be freed-up from the bureaucracy and processes that he believed was hampering R&D performance.

James wanted the project to run like a tournament or competition. Both teams would be given a budget, a team-leader and a set of explicit objectives in terms of the types of ideas and the number of ideas that were being sought. Crucially, they were also given a deadline for completion. A specific date was designated for the following July, some nine months hence, for the teams to come together at a meeting in the US at which they would be asked to present their ideas to James and his direct reports (the R&D Senior Leadership Team – or, SLT).

James asked his VP's to nominate people from their staff to be part of this initiative. Specifically, he was looking for the most creative people in the company to participate in this project with the objective of finding out what might happen if you put a bunch of top-class scientists together with very few constraints and a mandate to create radical ideas for which they have some passion. The tone and style of the proposal, which was still relatively embryonic, was very much aligned with the strategic direction being advocated by the company's CEO, Dr JP Garnier.

Not all the details had been thought entirely through and James's meeting then turned their attention to establishing what would need to happen for this programme to come to fruition. First, there was the issue of who to select for each team, what should be the ideal size of the teams; should they collaborate with each other or act independently? On this last question, James was adamant that he wanted the teams to compete against one another and so he did not want them to collaborate. Stan Lech (James's most senior direct report) nominated the two people he wanted as team-leaders. He nominated Nigel Grist (a senior research scientist who had been working on the Panadol analgesic brand) to head up the UK team and Scott Coapman (a project manager within the weight control product development team) to lead the US team.

His selection for UK leader generated considerable debate within the SLT with many people opposing the nomination. The reasons for their opposition are explained in greater detail in the pages that follow. However, Lech insisted and the decision to approach Grist and Coapman and offer them the opportunity to lead the teams was ratified by the SLT. Both Coapman and Grist were reporting in directly to members of the SLT; Grist was reporting to Dr Geoff Clarke, VP of R&D for Panadol and Coapman was reporting to Brendan Marken VP of R&D for Weight Control.

The SLT also scoped out the rules of engagement and ways of working that they envisaged for these teams within this initiative. After a great deal of discussion, the following approach was agreed. Sandy Lionetti, the head of Organisational Development, in an interview, explained what was agreed. The project would proceed and it would run along the following broad guidelines:

- They would empanel between 10-12 members in each of two teams – one in UK and one in the US.
- Insofar as they could, they would create the two teams equal in terms of the quality, technical expertise and experience of personnel.
- They would nominate the people to be assigned to the teams and they would choose those they believed to be the most creative in their groups.
- They would provide a budget of $250,000 to each team.
- The teams would not be required to provide any report on their progress during the lifetime (nine months) of the project.
- They concluded that team members be asked to allocate 20% of their time to the ISF project but the SLT would make no formal or official approaches to the line managers of the team members.

- Members would be expected to manage their own diaries and current projects in a way that would release the necessary time to engage in this project.

Lack of senior management interference was always supposed to be a hallmark of the project and this is evident in the five-slide deck that was used to brief the new team. Adding a little more detail to the background of the project, one of Stan Lech's R&D VP's, Teresa Layer explains her understanding of the project:

The first time we had a talent review process within the R&D community and we created a category of people that we felt were innovative. Described as innovate and the question that we discussed at talent review, well, if you've got a group of people that identify as innovators, why don't we utilise their skills in a different manner from classical projects? And so was really the idea that sort of stimulated the ISF. Where we said, you know these people are highly creative, highly energetic, have a lot to bring to the organisation. Why don't we, having identified them from talent review, which is the first part of the question "how do we find them?" Having selected them why don't we ask them to work on, dividing into two teams, identifying new ideas for the company through whatever creative processes they want to? To basically utilise with a completely open brief unconstrained by interference by senior management and to for a period of time work on that and report back with their ideas.

They then turned their attention to the the specific objectives of the project. Earlier in the meeting, there had been a portfolio review of the projects in the R&D pipeline and so they were able to suggest target objectives for the project teams. Each team would be given exactly the same objectives (Figure 5.2):

Figure 5.2: Slide from the ISF Briefing Pack

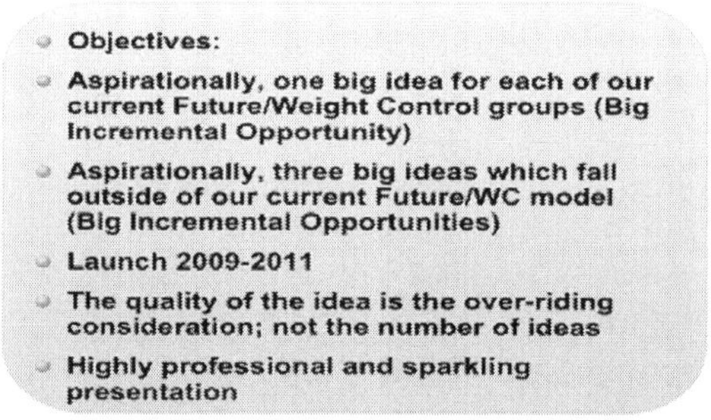

- Objectives:
- Aspirationally, one big idea for each of our current Future/Weight Control groups (Big Incremental Opportunity)
- Aspirationally, three big ideas which fall outside of our current Future/WC model (Big Incremental Opportunities)
- Launch 2009-2011
- The quality of the idea is the over-riding consideration; not the number of ideas
- Highly professional and sparkling presentation

Source: ISF Briefing Pack (November 2006).

Going through this point by point: The first bullet point asks for 5 ideas: one for each of the Future Group brands: Aquafresh, Sensodyne, Panadol, Smoking Cessation (Nicorette in US and NiQuitin in Europe) and Weight Control (Alli). The second bullet asks for three significant ideas (with a global sales potential of over $20m) in therapy or business areas outside the areas in which GSK was currently operating. The final three points are really qualifying the types of ideas that were being sought. Bullet three explains that priority will be given to ideas that would be capable of launch within three to five years. The company was not actively looking for anything with a longer-term time frame. Point four demonstrates that the SLT are not viewing this as 'a numbers game' but they are conscious of the fact that the quality of the ideas is the key measure and the number of ideas each team generates will not be a factor in considering the winners. Finally, point five demonstrates that they are looking for a polished presentation at the end of the nine months.

The targets for the project are very explicit and this slide is one of only five slides in the briefing deck. The preceding slide made it very clear what the measurable outputs were expected to be and the opening deck of the briefing set the overall context of the over-arching project objectives. It is worth noting that the issue of nomenclature around types of innovation is evident in this case study. The discussion within SLT referred to 'radical' innovation but when this got translated to the slide briefing deck for the teams, the word 'incremental', albeit preceded by the word 'big' was what appeared in the slides (see Figure 5.2).

Figure 5.3: Slide from ISF Briefing Pack

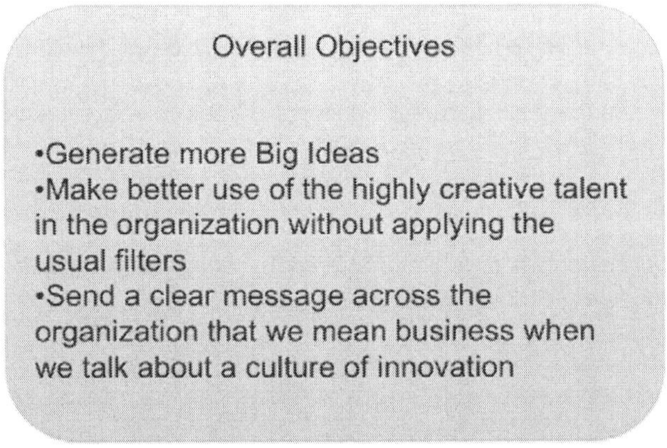

Overall Objectives

•Generate more Big Ideas
•Make better use of the highly creative talent in the organization without applying the usual filters
•Send a clear message across the organization that we mean business when we talk about a culture of innovation

Source: ISF Briefing Pack (November, 2006).

With the broad objectives and ways of working agreed, the next task was to approach the team leaders and get them on board with the project. Stan Lech, Ken James' second in command, was given the resposibility for engaging the team leaders and organising the two team briefings. The briefings were scheduled for November 2006; one in Surrey (UK) and one in New Jersey (US). The pages that follow contain the story of the UK team and their journey (Figure 5.4) from the initial briefing until they travelled to New York to present their ideas. After that, the US team's journey is described in similar detail. Much of the story is written in a series of direct quotations from each of the members as this is consistent with the semi-structured interviews conducted. Each quote you will read is either from a member of the UK team or one of the group of senior VP's who both set up and judged the outcome of the project.

5.4 Innovation Sans Frontiers - The UK Story

Figure 5.4: Timeline for the UK ISF Team

Timeline of events in UK ISF Team

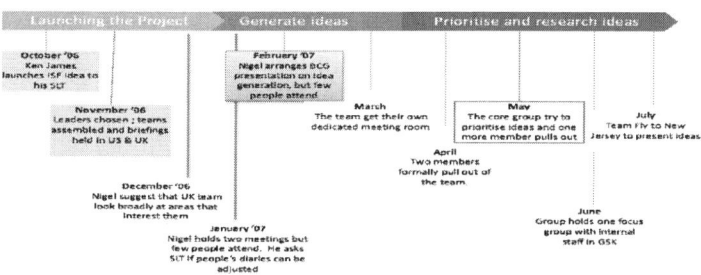

Source: Chronicle of the UK ISF Project produced for this study.

5.4.1 Selecting the Team Leader

Nigel Grist was selected as the leader of the UK Innovation Sans Frontiers Team in October 2006. Grist had been with the company seventeen years when this project was set up. He had always worked in R&D, first for

SmithKline Beecham (SB); then for GlaxoSmithKline (GSK) when the two companies merged in 2001. He had been involved in a number of functional areas within the R&D operation; New Product Development (NPD), tech licensing, external collaboration with third parties as part of the company's open innovation initiative; he had also worked on some of the company's key therapy areas such as oral care (the toothpaste and mouthwash ranges of Aquafresh, Sensodyne and Macleans) as well as the company's market-leading analgesic (painkiller) portfolio which includes the brands, Panadol, Solpadeine, Hedex and Beechams. Grist was a technical specialist within R&D. His role, up to this, did not require him to have any direct reports.

Grist was selected to lead the team; a role of considerable responsibility. Grist would be leading a team of twelve scientists over a nine-month period with a mandate to deliver specific innovation targets. The design of the overall project pitted his team against a team of similar size and similar levels of experience working in the US.

Grist describes the initial conversation with the SVP of R&D, Stan Lech, in which the ISF project was first outlined to him:

When I was approached about whether I would like to take on the challenge, the way that I was briefed initially was that we were looking to do something different. We were looking to give a group of scientists from within R&D total freedom.

Following Grist's installation as team leader, the attention of the senior leadership team then turned to filling the places of the team members. To get selected to participate in this team meant that each participant would have had to attract the attention of senior management specifically in the area of creativity. The head of R&D was looking for two very specific characteristics from all the members being assigned to this group. As Stan Lech recalls:

It was nothing to do with a popularity contest. We specifically picked people that we viewed as being at least two standard deviations from the norm, or maybe three, and it was behavioural driven...behavioural and track record, so we were very aware that we didn't want to put a bunch of weird and wacky people in the room that had no track records to delivery and deliver just a load of crap. So I used two filters; behavioural and delivery.

Specifically, in relation to the selection of the leader of the UK group, Lech reflects:

That was a bit of an experiment on my side, I think. I knew this going in, I knew Scott (the US team-leader) would bring the very structured approach, almost stage gating and I knew that Nigel (UK team-leader) didn't like process and so I knew he would be more of a free spirit around how it was all organised and that was a bit of an experiment itself.

Given that the entire project was something of an experiment (insofar as it had never been done before), the choice of UK team leader was really building an experiment within an experiment. Referring to Grist as 'allergic to process', Lech considered Grist to have considerable experience in various business areas and had a reputation for being creative and having an open management style.

That the selection of this particular leader was a managerial experiment was underlined by the opposition Lech faced from his own leadership team about his choice of Grist for the role of team leader. The VP of NPD, Brendan Marken, remembers registering his reservations to this assignment:

I had concerns that the UK team-leader hadn't got team-working skills; we had, in him, no leader, no co-ordinator and I raised strong objections to the point that my objection was noted. I was afraid we were setting them (the UK team) up for failure. My objection was noted but disregarded.

Geoff Clarke, also an R&D VP and on Lech's Leadership Team, had some misgivings about the selection of Grist as leader but he, apparently, chose not to voice them at the time. He considered Grist to be a good scientist and an experienced veteran of the GSK R&D environment but was concerned about his lack of team-management experience.

Nobody could be surprised about the lack of leadership from Nigel when he's never led a team in his life.

Grist knew nothing of the debate caused by his selection. He was attracted to the opportunity, because, as he saw it; it offered a welcome chance to work without constraints:

I saw it very much as a positive opportunity and something that would be a very interesting thing to do. So that was pretty straightforward; ... it was great to have the opportunity to do something completely without constraints.

He was to have a team of twelve for this project and they were each, in theory, to devote 20% of their time to this project.

5.4.2 Putting the Rest of the Team Together

The VP's of R&D then nominated people they felt would make a strong but particularly, a creative contribution to the team. The people thus chosen were nominated and assigned to the team; unlike Grist who had been personally approached and had the opportunity explained and offered to him (Figure 5.5).

His team members were simply assigned to the team, in some cases without the prior knowledge of their direct line manager.

Figure 5.5: Who's Who in the UK ISF Team

Who's who on the UK ISF Team

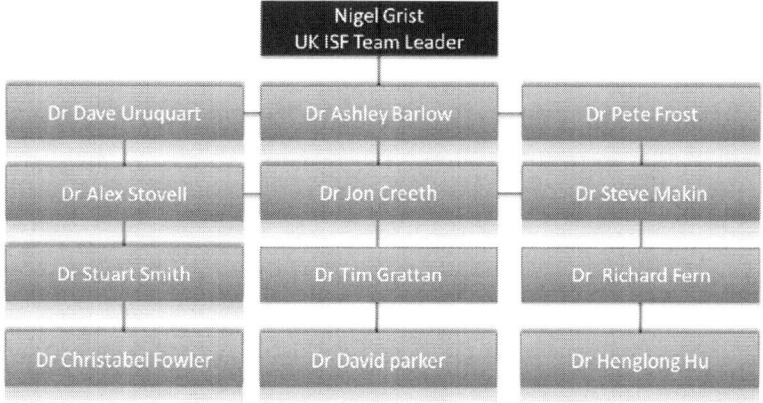

Source: Graphic developed for case study.

Note: The lighter boxes contain the names of the team members and the relative position of the boxes does not denote any hierarchy or reporting relationship. The members all had equal status within the team.

The other members of the UK team were nominated by top management within the overall R&D department. There were twelve members plus Grist. The composition, while intended to *'reflect a microcosm of the organisation'* - as Lech put it; comprised of eleven men and one woman. All were PhD scientists but eleven were from the UK with one from the US and one from China. They did reflect varying job grades within the organisation, ranging from a middle-manager (D-Grade) to a Vice President (Dr Stuart Smith).

The Vice President on the team, Dr Stuart Smith had his own reservations about the composition of the team; he also felt that there was not a sufficient blend of skills represented within it:

It felt to me that people have just put forward names of people who they considered to be creative or slightly wacky or disruptive or whatever and that doesn't necessarily make for the most creative team. I think actually having a balance of people might have been better, so actually potentially someone who was a bit more process orientated could have been useful for us I think. I think just putting a bunch of whacko's in the same room doesn't necessarily get people to feed off each other and lead to innovative outcomes.

Grist notes that, perhaps, from the very beginning, some team-members were less than enthusiastic about getting involved. He points out that in his own case (as noted above), he was approached personally and offered the opportunity to lead the team whereas many of the others were nominated by their managers and simply told of their involvement without having any choice in the matter.

I was approached, offered the opportunity, got the chance to buy in easy; (other) team members were pretty much told they were going to be part of this activity and I think it would pretty much be more of an encumbrance to the team members, at least initially, because it was just another thing they had to do amongst a crowded work schedule. So one of the recommendations I made is that people who are approached should just be approached about whether they would or wouldn't want to participate.'

The other UK Team-members were conscious of the kudos attached to having been selected for this team and acknowledged its connection with their prior track record for creativity and innovation. One senior R&D manager, Dr Ashley Barlow, attributes his selection to work he had done just before joining the ISF project:

I had been with the company three years when I was approached to be a member of ISF. The bulk of my research up to then loosely fitted under an innovation umbrella. I had been running dentin hypersensitivity clinical pain research. I had created different tools and different instruments – like brain imaging through to measuring oral health impacts. So, I suppose a lot of that work was very in tune with the innovation outputs needed by management – and I think this put my name into the picture.

Another member of the team, Dr Peter Frost had been closely involved in a very big, successful, international, new-product launch for Aquafresh and he saw his selection for ISF as part of the reward for this:

I was chosen because I came up with ISO Active because I had been involved in my previous job doing aerosols and furniture polishes. That led to project Mousse and Project Reindeer – both of which are launched now around the world.

This reference is to a (then) new product format under the Aquafresh brand which is gel-to-foam toothpaste. The product is, in essence, the collision of shaving gel format with toothpaste formulation to create a different and better mouthfeel experience when cleaning teeth. This idea had been developed by Frost, which he says occurred to him because of his prior history working with different liquid formats and aerosols. This variant now sells well over $100m for GSK and has been launched under various brand names throughout the world: Aquafresh, Sensodyne and Pronamel.

Dr Uruquart also reflects on the credentials in his background that may have brought his name to the attention of senior management:

Where I came from before GSK was Procter & Gamble in Newcastle working for four years in laundry detergents, so a lot of the chemistry was very, very similar. But obviously the ways of working from P&G to GSK were notably different and I brought a lot of that experience with me.

Another team-member, Dr Stovell was very flattered to be approached and he relished the opportunity to be involved:

Yes... it felt good to be picked for this team, I felt special and yes it was really nice to be working on something that felt had a real impact, was properly backed and supported from on high and I guess also it was quite nice to be working on stuff a bit behind closed doors, not everyone else could see exactly what we were up to, so I quite like that feel as well, doing something a bit subversive.

For the others, some, usually specific and notable, feature in their CV was responsible for their selection. They were chosen because of some specific experience or expertise in their track record that made their manager identify them as strong in the area of innovation.

5.4.3 The First Meeting to Launch the Project
November 2007

The kick off meeting was scheduled for Lech and his boss, Ken James to come to London, from their HQ in New Jersey, and outline to the team the parameters of the project; the key objectives, the resources available and the timelines and milestones along the way. Very quickly the new team members were appointed and invited to attend the briefing which was in November 2007.

The SLT's VP on Sensodyne, Teresa Layer attended the kick off session and noted that this message of how highly this group were valued by senior management; how they'd been identified as innovators, was strongly conveyed. The briefing also focussed on what was being asked for; what were the deliverables being demanded of these teams.   This was relatively tightly specified as per Figure 5.2.

Lech explained exactly what the senior management team were looking for:

*We said, look, we would love one to two big ideas for a **future group**, we'd love to have one breakthrough idea within one of the future groups and then we'd love to have – an additional idea; an incremental idea that might be worth up to £25m– and then we said on top of that we'd love to have £100 million breakthrough idea, which had nothing to do with the future group, which was new business opportunity. So that's what we told them to do.*

5.4.4 Getting Going with ISF

December 2007

The UK team then set to work. As it turned out, of the full 13 members, ten of the team members were co-located in the Weybridge R&D site in Surrey and the other three were based at the Lucozade factory site at Coleford some three hours drive away. Grist wanted to run the group with very little structure or process and he made this clear to the team members from the outset:

> One of the things that we tried to set out to do was not to have anything too process-driven, but to provide people with the opportunity to learn about new areas, find out new information, speak to new people, from which they could generate new threads. So the philosophy was that rather than have a process whereby you go through different stats to try and pull ideas together, is you create an environment where people have the opportunity to read about new areas, talk to experts in different areas, interact with different people such that they can generate threads for their own development. Now this was a little bit my way of thinking; it was shared, I would say, by half the team and I'd say the other half of the team really struggled with it.

Dr Stovell recalls how Grist ran the first meeting; what he suggested be done and how he hoped the team would perform in this challenge:

> Well the first thing to say I guess is that Nigel was quite keen not to have any process at all, he "de-processed" everything we did as much as he could, which he felt was the right thing to do. But in terms of how we approach the new ideas, the first thing he set us all to do was effectively to dig around and come up with any interesting areas, anything that people felt were places worth exploring, any interesting technology sort of things. So I guess he set us off on that before Christmas and then we had a meeting after Christmas for everyone to kind to bring along as to anything they'd found.

Grist himself points out that there were no commercial people on the team. The people one the team were scientists with a track record for creativity and that's what Grist wanted to amplify and explore with his group. He notes:

We took a slightly different view, which is that we were a group of scientists thrown together to come up with good opportunities. Now I would say there are broadly two areas where you can come up with huge wins in terms of new products. One is if you come up with a technical solution that is slightly different and can be positioned to an unmet or inadequately met consumer need. And that is what I would describe as a technically client-driven solution. Or the other way you can do it is if you just happen to strike on an insight, or a piece of consumer understanding, which enables you to make something of what you have already got available in a light that hasn't previously been recognised. As a group of scientists I felt our role was very much to come up with new technical solutions. Things that would at least be worth exploring from a technical point of view.

So I felt that we should at least stick with our scientific heritage, try and look at science in a different way, come up with novel technical solutions rather than trying to pretend that we were commercial folk when we weren't. So the thrust of what we were trying to do was very much get into the science, look at how you might use pieces of science that have been applied in one area translated to another area.

This distinction (between the primacy of science over marketing in how he approached the ISF challenge) is a key one in the narrative. Implicit in Grist's philosophy is that he wanted to focus on science or technology-driven ideas; to use the experience and expertise of his group of scientists to push the boundaries of new scientific knowledge. This is a declaration that he is not going to use the market and consumer research as his guiding star; he is favouring the 'technology-push' approach to innovation ahead of 'market or customer-centred' innovation. He is both very clear and very considered about this perspective, he further states:

So part of the experiment was very much to isolate scientists and see what they came up with and we felt in that context it was worth us trying to focus on novel applications of science.

Grist gave his people a mandate to go out and explore technologies and therapy areas in that attracted their interest. He wanted them to go out and talk to specialists, academics, opinion-formers and to chase down any potential opportunities in areas in which they had some interest or connection. His commitment to *'the science heritage'* reflects his view that his team were talented and experienced scientists and that his role was to leverage this strength rather than dilute it through dabbling in marketing areas. He was also committed to making sure his team were networking with external experts, universities, inventors and attending conferences to explore new, appropriate scientific areas. He saw his management style and role as that of a facilitator:

We were inviting the external experts; we were highlighting to each other interest in sources of new scientific development and trying to create an environment where people would explore new areas and find out something they thought was of interest and then develop threads. I don't know how much sense that makes, but it's the way that I'd seen the most creative people operate who I'd worked with within the organisation. So very much individuals taking responsibility for developing their own threads of research, and interacting with other people to maximise where they might go with that.

In terms of idea generation, Grist did not provide any particular guidelines except that he asked people to look to their own technical interests to see if they could find ideas of possible commercial potential to begin to explore. In practice, this translated into some of the team looking at certain healthcare therapy areas, such as, in the case of Dr Stovell; sleep. In the case of sleep, Dr Stovell contacted the National Sleep Laboratory in Guildford University and arranged to meet them and talk about possible gaps in treatment or possible areas for self-medication. He brought their ideas and feedback back to Weybridge and he began to develop potential ideas to launch in the market for sleep aids. Other members of the team began the process by examining areas in which they had a specific interest and/or experience. Stuart Smith began by examining opportunities in gum health.

While, for some, there was great emancipation in this open licence to go out and experiment with new areas and ideas; others would have wished for more structure to be introduced. One member (Dr Jon Creeth) recalled:

The UK approach was chaotic with process largely absent. It was like a playground where you could do exactly what you wanted.

Dr Ashley Barlow, one of the team-members located at the Coleford site suggested:

I think some people liked our team's approach however, we were all smart enough to realise that unless you have some structure in place there was a risk that the chaos would have very little output. So most of us wanted gates to aim at and deadlines to hit.

Dr Peter Frost agreed:

Nigel used very little framework for running the team. He never set the scene but allowed everything to be completely flexible. While (the) innovation (process) needs chaos – maybe the people don't and a bunch of scientists generally want some structure. I think it was a little bit chaotic and over the top. I could have accepted it if at some stage, he intervened with some structure and showed us a path towards a goal.

Frost's perspective was shared by Dr Jonathan Creeth who, similarly, found the absence of structure for the project quite intimidating and something of a barrier to his full involvement:

But it was quite intimidating because it was so open ended. If someone starts something off and need the edges smoothed – then I can help out. But when you get the totally blank piece of paper – which we never get -

*and the only hard thing on it was a deadline for delivery. I thought that
was intimidating right from the start.*

A few in the group, though, were self-starters and they relished the freedom to
go out and talk to experts in different areas, but nonetheless, they
acknowledged that some of their fellow team-members were struggling to make
a strong contribution precisely because of the lack of structure. Dr David
Uruquart had joined GSK from P&G and he spotted this difficulty straightaway:

*What I would say is this was a very scattergun approach. Nigel was very,
very keen that he wanted to go as wide as possible and that he didn't want
to put any process in because he felt at that stage in the idea gathering
point that he didn't want to put any barriers on what we looked at.
Personally, I think that helped some people: but for many others, I think it
hindered their innovation capability because being given a completely
blank sheet of paper almost moved them into torpor.*

But the team had been asked before Christmas to look broadly into areas of science in which there may be potential opportunity and to come back in January with some early stage ideas of where further exploration might be warranted. However, up to half of the team failed to come up with anything once the team reconvened in January. Their first working meeting took place in Weybridge where the Coleford-based team were allowed to join on a videoconferencing line. Dr Stovell recalls:

I think we had a telecom with Coleford for that one; a video con in the formal VTC unit. As I remember, we didn't have everyone there and I would guess about only 50% to 60% of the team had actually done anything at that point.

February 2007

Grist arranged three consecutive team meetings in January and February. The Weybridge site was the venue with the Coleford contingent dialling in on a conference call. The meetings were intended for people to share their ideas but it quickly became apparent that some people just hadn't got started and came with no progress to report or no ideas to share. The meetings were not seen as a success by the group and it also became apparent that less and less of the team were showing up and that some people were gradually, informally but unmistakably pulling away from the project.

Progress was far slower than Grist intended or had hoped for. David Uruquart felt that Grist ought to have shown more responsiveness when he observed that some of his team experiencing difficulty in making the type of contribution he had been expecting from them:

I think he was so married to his idea of how it would work that I think when things didn't work he was relatively inflexible, and I think I would have liked to have seen, for instance, his 'I don't want any process' became a mantra that despite the fact that we'd had about three different meetings in a row where frankly there'd been very little progress made, and people were obviously struggling to connect because they weren't seeing it as he was, I think he could have been more flexible with his approach. And I think he could have understood more the issues people who were in the team were having and tried to make adaptations to his style in order to do that.

Grist knew the team was experiencing difficulty; he felt that their progress was slow. But, as he had no schedule or plan, he had nothing to compare it against. He began to become a little disillusioned with the attitude and performance of his team. He reflects, in this quote, on how his prior assumptions about the innovativeness of his people may have been mistaken:

So, I had started from the position that people who are considered innovative should all be self-starters within the organisation. So that, as it turns out, was an assumption too far, but I was working on the basis that everybody was at least individually driven to explore new avenues; and I think some people are individually driven, but not necessarily to explore new areas, more just the ones they're already familiar with.

Grist made his first use of the budgeted resources, in March 2007, by hiring a specialist company called Bufton Consulting to come to one of his team meetings in Weybridge. Bufton Consulting are an Open Innovation Intermediary company. GSK had worked with Bufton Consulting before, they had used their services as technology scouts in the past. Grist had privately shared with them the ISF brief and they had examined it from the standpoint of their own commercial contacts; a group which includes inventors, research institutes, universities and small start-up technology companies. In this way, Grist farmed out the task to get a commercial view on how a professional consultancy might approach it. He also wanted to broaden the external contact base or network for his team and intended that Bufton Consulting might broker useful introductions for his team members. This did have the desired effect as Bufton steered the UK ISF team towards some of the ideas that would eventually make it onto the team's final presentation.

Bufton had some proposals for the ISF team in terms of companies who were developing new technologies in adjacent healthcare platforms for GSK. Two directors from Bufton Consulting came into GSK Weybridge to present their ideas and their contacts to the UK team. However, less than half of Grist's team turned up for the meeting. Nevertheless, this did help to get some people going and generating new ideas but it did not succeed in bringing the group closer together. The Coleford group did not attend this session.

Although a number of team members reported that they were beginning to struggle and had begun to seek direction; the provision of such specific management direction ran counter to Grist's view of how the project ought to run. He felt that if people needed this type of leadership, then they weren't the right people for this team. He was unswerving and emphatic on this point.

If I am going to be blunt about it, the people who were the right people on the team were the ones who felt energised and liberated by the freedom to explore where they wanted to go. The ones who were the wrong people on the team were the people who wanted basically spoon-feeding, directing and task allocation, and all the rest of it. So here you probably see my values coming through, my personal values; I have a view that if someone is a scientist and they are chosen as an innovator, then they should be self-starters and they should feel motivated and liberated by the opportunity to start exploring different areas. So for example, from the scientist's point of view you are told 'go innovate', you've got a blank sheet of paper, find about a disease condition that is affecting a lot of people, which there does not appear to be an unmet need; go and read about it; go talk to some experts and see where it leads you.

So that is very much my philosophy, so I think the answer to your question is the right people who were on the team did feel motivated and liberated

by having total freedom in an environment where there was also some
funding available and support from senior management to do it.
Dr Smith, the most senior member of the team in age, grade and experience, expressed the view that the absence of a very tight structure was helpful in the case of the UK team; he refers to the fact that the team had a lot of 'big personalities' and that a tight management approach may not have been productive. But, this was Grist's dilemma; the 'big personalities' might very well have resisted too much structure, but without it, they had a better opportunity to dominate proceedings:

> *I think if someone had tried to drive it with too rigid a process, I think we*
> *would have lost the creativity side. There were some strong personalities*
> *and a range of very different personalities on the team and I think there*
> *potentially, some people got a too greater share of voice.*

Grist's philosophy was by this stage, certainly at odds with the declared needs and wants of the majority of his team. But Dr Frost points out that the team themselves could have come to a collective decision about the type and level of management structure they actually needed. He believes the team could have discussed this and made appropriate representation to Grist. But, they didn't.

> *In a situation like this, you can't win. Maybe the way to do it is to allow the*
> *team decide how much structure they want. You should ask not how is*
> *the leader going to run the group but ask how the team can lead the team*
> *– what do the group want in terms of structure and leadership.*

5.4.5 Difficulties Persist in UK Group

March 2007

The project was structured on the assumption that everyone would devote 20% of his or her time to working on it. However, no provision was made to adjust people's objectives or lighten their workload accordingly. This became an increasingly difficult issue for team-members and was especially magnified as the team itself had not really evolved into a cohesive group and hence some people were prepared to abandon the ISF group if they felt their 'day-job' was suffering. By April, two members had formally quit the team with another five largely disengaged (insofar as they were not contributing anything) but still formally, or technically remaining members.

Grist found this an acutely difficult issue to deal with and he struggled to persuade people to stick with the project. One of the exacerbating issues was that although the individual team members had been approached by senior management to let them know they were participating, in many cases that person's direct line-manager had not been similarly approached and this lack of foresight resulted in some unnecessary tensions occurring. Grist describes it:

> The only thing I would say is the time issue was a huge one, which is that the ideal was that people spent 20% of their time on this activity, but it was pretty clear that none of their day job was going to be allowed to disappear as a part of that equation; so people were left often doing it in the evenings. That would be my only question about the level of support, certainly in terms of allowing us to get on with it, leaving us as an autonomous group. I thought that as part of the experiment, that was all excellent. My only query is around the time.

Grist raised this issue with the Senior Leadership Team back in November when he foresaw it becoming an issue; asking them if any provision could be make to clear some people's diaries and alter their other project timelines but was not given any leeway in this regard:

> In fact that question was asked at the beginning and it was made pretty clear that the day job was not going to go away; to some extent I could see that left a few people feeling somewhat burdened from the off. In fact two or three people pretty much dropped out.

An interesting feature of the overall constraint on the team's time availability was that, generally, the higher up people were in the organisation, the more pressing were the routine demands on their time and the less likely they were to be able to contribute to a project like ISF. This left an attractive opening for the younger, more junior members of the team to 'punch above their weight' and to advance their own ideas more easily.

Dr Alex Stovell, the team's youngest and most junior member recalls:

I think ability to commit a significant amount of time would be one of the biggest barriers. I guess I was lucky in as much as I was one of the more junior members of the team, probably the most junior actually, and I suspect that the amount of time that you could free up to work on this was probably related to grade to a certain extent. So the higher up people were, the harder it was for them to dedicate a sufficient amount of time to it. That being said, I think I would have made time for this just because it was great, I would have found the time. So I think a certain amount is down to individual's attitudes to it as well – they just drifted in and out; some people deserted the team because of the time issue and their other priorities.

At the other end of the spectrum in terms of experience and age; Dr David Parker who was one of the members located at the Coleford site where they handle R&D and manufacturing for Lucozade, Horlicks and Ribena, had a different view of the time availability issue and he found it a genuine barrier:

Time was a key barrier – 20% of your time was just not feasible. Great innovation doesn't come easily out of thin air – it's a combination of great curiosity and a great deal of thinking. One day a week was the wrong way to start it off. You can't come up with a great idea just because today is 'innovation day....

In fact, Dr Parker was one of the first to have to abandon the project purely because of his inability to find the time to participate:

I withdrew from ISF very early - never really got into it - due to lack of space in my schedule.

For Grist, this was the key issue he faced at the early stages; people were feeling the pressure of contributing time to ISF while still being asked to deliver a full day's work in their 'day job'. He would have liked to see some management support offered to him in this area but none was forthcoming.

The gist of it is that there was some initial resistance from people who, at a time when there was a lot of pressure on the organisation, and they were already working relatively long hours, saw it as another thing they were being asked to do without giving up any of their existing work time, so it became a little bit of an encumbrance. I felt some of that could have been avoided just by inviting them and selling it as an opportunity a little bit more, rather than just telling them they were participating.

Like most organisations, GSK is rarely static, and this initiative was taking place against a backdrop of significant change in the company. The R&D Group generally were going through a transformational change process where they were re-branding themselves as 'Radical R&D' and this entailed new job descriptions and new roles for certain people. Dr Creeth remembers:

Time was a huge problem. I was in a new job and was thrown in the deep end in a new job and to do this well it required more time. It was just at the wrong level; the worst level – if it was more than 20% we could have got more involved – if it was less than 20% then you could fit it in.

David Uruquart sums it up:

So frankly there were a number of people who dropped during the way because they couldn't commit any time to it, and there were a number of others that did actually go to the end, but frankly committed hardly any time to it whatsoever. People's allocation of time to the initiative was inversely proportional to their seniority in the organisation.

Those located off the principal R&D Weybridge site seemed to have more difficulty in staying the course of the project. They may have felt the lack of structure and process more acutely than the others who, at least, had proximity to each other to fall back on when they wanted to discuss and review ideas. Dr Barlow, in April, reluctantly, had to quit the team:

I found the time required became increasingly difficult to deliver. I had to stand down from the team because my other projects were suffering badly. So, I had to stand down or otherwise I would have been a ghost member of the team.

The other members also noticed how difficult it was to sustain the appropriate environment and atmosphere for creativity and innovation when you were only connected on a conference call and so were not surprised when the majority of the Coleford group eventually disengaged from the team (one left formally in February and another in April). Dr Smith recalls:

And I would say another challenge the UK team had was having to do pretty much everything by teleconference because we rightly had the representatives from the nutritional business in Coleford and I think, that does impact the dynamics of any meeting if you consider doing things only by phone.

Dr Stuart Smith also broached the issue of time as a barrier to people's involvement:

I think 20% of the time was the intended bit. But I think the challenge which became apparent is 20% of time was not made, so it was one of those things you spent 20% of your time, but you have to do everything else as well. And I think that impacted heavily on the team because everyone is so busy it's very hard to find 20% of anyone's time unless you actually reprioritise and cut things out.

Dr Frost also mentions the time issue as an inhibiting factor to the team's progress:

> *Biggest flaw was that we didn't have time to do it. The one big learning is that you allow people enough time or you don't do this at all. Many people couldn't commit to the meetings and didn't contribute as they might have.*

Only one of the team members, Dr Stovell, questions the authenticity of the issue of time availability; he suggests that some people may have been using it as an excuse to explain their own lack of interest, enthusiasm and involvement:

> *I'm not totally convinced that these people wanted to free up their time, they were using it as an excuse, it's kind of the feeling I got in certain quarters.*

5.4.6 Channelling the Team's Work
April 2007

But meanwhile, some positive steps were being taken and ideas were being generated and incubated by the group. Despite the few 'casualties' along the way, a core team informally evolved to become the ideas engine for this team. Dr Stovell, who had the time to devote to the project, explains:

> *Of course, there were others who didn't turn up to many meetings. But there were, I guess sort of three, four, five of us who did as much as we could, turned up for as many meetings as we could and really got behind it. So we kind of had a bit of a sub-unit that was working well together, specifically I guess between me and David just because we always work well together and have done before and we have got the common sort innovation language. So that was a natural pairing that worked well and we carried a few others with us. And we did deliberately jump in on the sleep area to try and ......drill into the sleep (therapy) area as a new area and we thought at the time that this would be a good example to others as to how they might drill into any other new areas because they could do a similar thing. It didn't actually happen, nobody else got to grips with any area in the same way that we did with sleep.*

While some people were used to working on teams or sub teams and were expecting to be paired with a partner, or a small group, to work on something; they were disappointed as Grist allowed things to evolve organically. This meant that for most people, although nominally on a team, they were actually working alone. Dr Creeth recalls:

> *I think we were united in our approach but we weren't like a coherent team. I thought we'd divide ideas up between two's and three's and some of us would work in sub-teams. Only people who had overlapping interests worked together – mainly we all shot into the distance and worked apart from each other.*

There was also a sense that just a few of the original 13 were doing all of the team's work while the others disengaged to varying degrees. Dr Uruquart recalls the lack of engagement by many of the team:

> *of the twelve there were probably three people that spent the required amount of time in it, and we are...I can be counted as one of those, we spent our time on it additional to the day job.*

Dr Stovell also notes:

> *The entire UK group never actually met once during the entire period...*

Dr Smith also has this recollection that the team was more a collection of individuals with no actual time when they all met together:

> *It was very difficult for the team to ever actually get to meet in its entirety I don't know whether we ever actually had everybody present at the meetings.*

5.4.7 Creating the Team Spirit and Psychological Safety

For those that did commit and participate, Grist was very open about the ideas they could pursue and he created a very positive environment for people to volunteer new and fragmentary ideas. All members, who attended, commended the atmosphere in which the meetings were conducted; people felt encouraged to contribute raw, incomplete ideas without any sense of self-consciousness. Dr Frost remembers the brainstorming meetings with some fondness:

> *On the positive side, it was quite sparky, quite innovative, we used our networks to generate ideas that we were interested in - and a small number of us met up every two weeks or so to download all our ideas. It was a great notion getting creative people together to build on people's ideas. It was great fun – I loved it.*

Grist made it clear that he was open to radical ideas and this stimulated some members of the team to deliver them. Dr Creeth recalls:

> *There was a good climate built with the team – there was recognition that if it didn't sound wacky in the first place, it probably wasn't any good!*

But there were some within the team who weren't sharing all their information and who, some people suspected, were using this group to incubate ideas which they intended to deploy elsewhere or within some different group. Dr Frost was sceptical about the participation of some:

> *Some people, even in meetings, were quite guarded - some of them were (working) on other initiatives to do with innovation and I felt they were keeping some of the ideas for another purpose.*

5.4.8 Choosing which Ideas would make the Cut
April 2007

The team continued to generate and develop ideas, albeit mostly as individuals rather as a collaborative group or team. By now, over half the project time had elapsed and some among the team were beginning to get anxious about starting to focus on the very best ideas that they could present at the final New York review. This involved bringing people together and trying to get consensus around which were the ideas of highest potential.

But Grist continued to see his role as facilitator, just to encourage people to follow the ideas for which they had a passion rather than to impose any formal structure on getting it done:

A huge amount of ideas were generated often by individuals or small groups. We'd then take a look at it as a team and point out what we thought the merits or challenges of the different ideas would be. And once you start sharing discussions amongst a broader group, it became more obvious what would or wouldn't necessarily be the best opportunities. So it was done a little by consensus, but one of the guiding principles was that if somebody felt really strongly about an idea and they were prepared to present it to senior management and champion it then they would get the opportunity to do so.

Some of the team members expected more prescriptive and directional intervention. Dr Frost recalls:

We had meetings which ought to have been moving the ideas forward but instead ended up being mired in the minutiae of individual ideas because there was no formal process to select the high-potential ones.

Dr Barlow notes that personalities played a big part in which ideas were pursued within the group. He notes that the personality and position of the originator, rather than the inherent quality of the idea, may have had more influence on which ideas were chosen to make the final presentation:

There were ideas that were very personal to members of the group and largely went through purely on the strength of the passion the individual had for the idea. Personalities came into play. Unless there was an obvious fundamental flaw, people were happy to see those ideas carried forward especially if those passionate individuals were happy to do the groundwork.

Dr Frost concurs with the point that the strength of the personality of the originator played a part at least equal to the strength of the idea itself in getting ideas supported:

This is where the system failed – because we had lots of very strong ideas. Some of them made it through because of the power of the personality rather than the strength of the idea itself.

Dr Smith recalls that the team's process for screening ideas and ranking the ones they wanted to focus on was hampered by the fact that they didn't apply any dedicated, customised screening process. He notes that they really 'chatted through' the options and ideas:

In teasing out the positioning and the difference between something that might be huge and something that might be "a lemon". I think a screening process would have been a good discipline to go through if it was done properly; I think what we did was to use a bit of experience and ... I think if we were going to have gone a step further than that we would have needed to do a much more in-depth analysis; I think there is probably a question was what the end points since we chatted through. I think it probably, looking back at it, it's probably not that clear what the expected end point of the process was.

Dr Barlow focuses on how the absence of rigid structures allowed people pursue the ideas for which they had most enthusiasm;

No formal process for ranking or screening ideas; we just went with what people thought were good ideas at the start.

Grist's view was simply that if one of his team had a strong passion for an idea, then he would support them to continue to run with it:

Absolutely, because that was one of the challenges when you come to do the idea screening, inevitably people have a lot of ideas and sometimes you have to do it by majority voting and that can leave a very sour taste in people's palates if they have a real passion for something and other people don't share it. So the acid test that we applied was that if you're prepared to stand in front of senior managers for three or four minutes and justify why it's such a good idea, you get the opportunity to do it.

Dr Smith, the VP (most senior in terms of rank) among the group, thought that it was quite a positive thing that people with passion for ideas should be allowed to continue to make the running with them until they ran out of steam. He reflects:

Whilst we did have some that we were able to kill off for a variety of reasons, I would say that if somebody had genuine heart and enthusiasm for an idea, they were allowed to run with it which actually I think is the right way of doing it. I think that for this sort of process, the enthusiasm is more important than having to put the stickers on the chart and decide which will be best, because remember we're doing it from a scientific perspective. So some of the areas of science would not have been strengths of everyone on the team, so there had to be an element of trust, if someone felt this was an area that they were expert in and felt it was a good area to explore and that would carry on.

5.4.9 Rising to the Challenge

May 2007

This was a difficult time in the life of the project. Grist was becoming a little more disenchanted with some of his team and some among his team were becoming equally disenchanted with him. The tension was amplified because the deadline was looming and although the freedom to explore novel avenues was a very liberating factor in the beginning of the project; now that the finish line was coming into view, the team began to experience some concern about whether they had been going on the right direction.

Grist began to wonder about just how innovative his team actually were:

The working styles and personal capabilities of the people on the team were hugely variable. Some people were not innovative. I would say that others have a track record for being hugely innovative and you start off with an open mind, clearly, and you assume everyone's innovative. By the end of the project I was wondering how some people had been identified as innovative, when in my view they weren't.

Grist's view remained that the people who were truly innovative would rise to the challenge and their natural creativity would emerge regardless of how the team was managed. But, some found this a very challenging environment in which to work. The total absence of direction continued to intimidate many team-members; Dr Stovell recalls:

Nigel's style did not suit everybody. There were people there I think who just needed to be told what to do which is in some ways surprising from a team that was supposedly selected for creativity. Many would have benefited from more structure – they wasted lots of time being rudderless.

Some people had an affinity or natural inclination towards, either certain types of innovation projects or else specific dimensions of the innovation process. Dr Creeth raises this point:

Normally, we plan our projects and you have an outline in your head. In new product development you don't come up with the new ideas – you embellish them. So, it's very rare to have a completely blank page.

Dr Parker, one of the three Coleford-based members of the team was also frustrated with the leader's approach:

The leader didn't provide sufficient clarity of purpose and then let individuals get on with it....there was no clear leadership. We 'faffed around' a lot for the initial phase and then scrabbled around at the end....

Dr Creeth concurs that a formal, open idea-screening process was really neglected and that people had the licence to carry through ideas purely on the grounds that they felt a sufficient personal interest in them. Hence, there was some doubt within the team that the very best ideas were being progressed.

> Most of the ones that died were ones that people weren't sufficiently bothered about. The palette of ideas was formed by what people wanted to do rather than a collective view from the entire team.

Grist was aware that certain people were disengaging and he decided that rather than try to convince them to either stay with the project or return to it, that instead, he would focus his attention on the core group who were really committed and were contributing:

> As time went on, it became increasingly clear to me that if we were going to deliver anything as a group, trying to get the individuals involved who were not contributing that much was just wasting my effort and time. I know that, so I switched from trying to support everyone, encourage everyone, to actually working with those who were committed, because that was the way we were going to deliver the most by the deadline. So they got to a point halfway through where, if people were expressing a reluctance to contribute for whatever reason and said to me, "look I don't think I can do this", I said, "I'd agree, yes, let's get things re-arranged" because it was actually draining my time and energy trying to get people into the fold if they weren't really there.

June 2007

Even at this late stage of the project, time and momentum was slipping because many of the team members were either looking for more direction or were looking for a more collaborative approach where they might get paired with someone or placed on a sub team to make their contribution that way. At this point, Grist decided to spend a little more of his budget and he brought in a professional market research company; Frank Research. Frank Research is a company which is used regularly by the GSK organisation for qualitative research. Based in Oxford, they are experts in small scale, impressionistic, qualitative research. Grist commissioned them to expose some of the group's ideas to potential consumers so that he could get an objective steer on which were the lead ideas and had the most appeal.

Hwever, Grist had left it a little late to call Frank Research in to perform the sort of focus groups that they would normally run in this type of project. Frank's director of client service, Hugh Shelton pointed out that they wouldn't have sufficient time to recruit the research respondents, for focus groups, externally. A solution was found through which the respondents and focus group participants were found internally within GSK, from among staff on one of the sites, and the research was run during the course of the day within the GSK office environment.

Dr Uruquart, describes how this element of the process worked:

> *And then we took the whole range of ideas through essentially an internal consumer group that we run at GSK House to start identifying which ideas triggered with people and which ones worked. And then also similarly at the same time we were looking technically for evidence to say which ones we think technically would work and then we married those together to come up with our lead ideas.*

Grist notes that, for him, this customer interaction was not the biggest priority for the project and he deliberately left it till last thing on the team's agenda:

We didn't want to neglect consumers completely; because we recognised they're the end customer. We just felt that in theory at least the onus was on us as a group of scientists; ... So part of the experiment was very much to isolate scientists and see what they came up with and we felt in that context it was worth us trying to focus on novel applications of science rather than looking at customer wants.

Then, Grist got a lucky break when someone who had exited the team and not contributed for 6 months, arrived back to help out towards the end. Grist remembers:

Another team member who had unofficially left, and they came back into the fold with three or four weeks left, were hugely beneficial to the team, but were not in a position to contribute over six months.

Grist also became more proactive in managing the contributions of the team as a group. He had noticed that little more than half the team were contributing and he decided to re-engineer the team by formally asking those that were 'drifting in and out' to formally leave the project. He explains why he wanted to weed out the people whose contribution was negligible so that they wouldn't dilute or contaminate the enthusiasm and energy of the rest of the group:

So one of the things I observed is that if you've got half a team committed and half a team that isn't - I'm using half as a simplistic estimation - then actually that half that's committed will be far more motivated and focussed if you get rid of the half that isn't. So it's the old one about if you've got people not really involved, drifting in and out and you've got different levels of commitment to the cause, actually it's hugely beneficial for the team to jettison the ones that aren't committed.

Once Grist had stripped the team down to the core, active group, they began to refine their ideas and decide which ones they wanted to present. The team reported that they now worked with more agility and purpose and they started to craft their final presentation. Dr Smith recalls:

> When we did the presentation I think there was a great feeling of team spirit when the presentation was being put together. Before that I think there was probably a tendency for it to be a bunch of individuals who have been tasked with something.

Grist wanted the presentation to be 'slick and professional'. With the team, he discussed having music and some strong visual stimulus to accompany the presentation. His team started to assemble images of nature and images of innovation in science (e.g. the Apollo space craft) which they would use during the presentation. Some images they had professionally printed onto exhibition stands and banners for deployment in the room in which the presentation was to take place.

Dr Stovell was closely involved in the preparation for the big presentation in New Jersey:

> I think it was a three hour presentation to the SLT in Parsippany, so I think we had the morning and then the US team had the afternoon, just to be sympathetic to our jet lag. So I think we had eight, if I remember rightly, of the thirteen strong team who managed to make it for that and we mixed it up, we had presentations, we had table demos, we had little factoids chucked in because we had just this huge volume of interesting stuff that we wanted to give a flavour of, so we just chucked in little bits and pieces here and there, had some posters as well.

On arrival in the US, the UK team decorated the conference room in which the presentation was to be made with vibrant and colourful images of nature. They also printed giant posters of people associated with scientific invention like

Einstein and with artistic invention like, Dali. They used powerpoint as the medium for their presentation and some of the ideas had prototypes which the team demonstrated on small tables on the perimeter of the room. They presented 27 ideas in total and these were grouped by theme; such as ideas around sleep (therapy) or ideas around skincare. Presenting everything they had is consistent with their open and fluid approach. But, later, Dr Uruquart, thought that this had been a mistake on their part as he thought presenting so many ideas made it difficult for the SLT (the judging panel) to focus on the specific, individual ones that might have real potential for the organisation.

Yes I think if we have been able to get some more concrete numbers around some of our better ideas we could have maybe cut 25 down to 15 and then they would have been presented in a stronger fashion and got more purchase within the organisation, because of that if we had a bit of better evaluation on them rather than just present pretty much everything that we had if we got to that stage.

5.5   ISF – The US Experience

November 2006

Scott Coapman was chosen as the leader of the US project team for ISF. At the time of the project, he had been working for GSK for exactly 10 years. Prior to that, he worked for six years with P&G also within R&D. Within P&G, he had been responsible for early stage qualitative research on some of their new business ideas. In GSK, Coapman was a (Global) Project Manager within R&D. This is a role which is different from a technical expert. The role of project leader brings the discipline of project management to bear on all elements of the project and works alongside a technical lead on most projects to make sure all the complex elements of the project move along with sufficient pace. Coapman describes:

It has the same remit of project managers overseeing the project, but now it was no longer just the technical side, it was the R&D work streams, which involved the regulatory, the clinical, the safety; it was the commercial work streams which included any type of consumer research, advertising, development schedules, ultimately any of the commercial activities that were required to get the product from idea to the shelves of Walmart, with on-air advertising.

In terms of career development, Coapman had voiced a desire to work with the commercial teams alongside the technical R&D side as he had done in P&G and he made this wish known to his manager, who at the time was Ken James (later to become the company's President of R&D and architect of the ISF initiative). The opportunity presented itself in the form of a marketing secondment to the company's US operating HQ in Pittsburgh. Coapman recalls:

Well, it took about five or six years before the opportunity presented itself and it nicely did with an opportunity in smoking control Pittsburgh NPD marketing group. And it was really doing just the sort of stuff that I thoroughly enjoy doing whilst at P&G. So doing some of the upfront consumer research, writing concepts, going out talking with consumers and focus groups. Fielding some quantitative concept studies, working back with the folks in Parsippany to make sure that these concepts weren't promising something that just technically was a pipedream. So I was actually in Pittsburgh for nine months on that.

Following his selection as team leader for ISF's US group, Coapman described the target outputs were for his group:

And it was all around the concept of let's get a group of R&D people and it was intentionally made just R&D folks for this round of the ISF and let's give them an operating budget of around a quarter of a million Dollars and let's give them nine months of time; and the brief was essentially, go for it come up with some ideas, some new product ideas and report back to us in nine months. And it was really pretty much that open. But they essentially said aspirationally the output they would like to see 1) good size idea for each of the future team brands and then aspirationally three new ideas, new product ideas that were outside of any of the therapeutic categories that we currently operated in.

He also described the scale of the ideas that were being sought:

The other thing that they wanted to see was for any of these ideas, they didn't want things like just simple line extensions, like a new flavour change to Aquafresh but really something that was rather big. And they said you know for purposes of trying to gauge a big idea, the thought around £20 million incremental was thrown out as something to target for.

He was conscious of proposing ideas which had a chance of being brought to market within a time frame of three or four years:

Another area that they asked us to try to ring-fence the ideas too, are things that could be brought to market by 2011. So, we found through our ideation, through the whole ISF project, we found some ideas that we thought met the big sales volume criteria, but since they were FDA (requiring extensive clinical trials) type projects and such we didn't see that they could make it by 2011 so those were parked.'

As with Grist's team, Scott's team were selected by the SLT and he did not have any influence on who was chosen. Coapman was given a list of his team members (Figure 5.6):

Figure 5.6: Who's Who in the US ISF Team

Who's who on the US ISF Team

Source: Case study interview notes.

Note: The lighter boxes contain the names of the team members and the relative position of the boxes does not denote any hierarchy or reporting relationship. The members all had equal status within the team.

The US team were only 11 in total whereas the UK team started with 13 members. However, there is considerably more diversity in the US group with three women and more international representation, especially from Asia. The US team were also all based in one location; the company's US R&D facility in Parsippany in New Jersey.

Stan Lech arranged a briefing in the US R&D Headquarters in Parsippany where all of the team members were based and where Lech and most of the SLT were also based. The same slide presentation was used (as had been shown in the UK earlier the same week) and this constituted the start point for the US team. Within that first week, Coapman got the entire team together for a meeting. Coapman had a clear plan for how the team were going to accomplish what was expected of them:

When I kicked off the meeting, I had a very clear vision in my mind of what we wanted to be in a position to present to SLT 9 months later. And it wasn't just, you know, some ideas that have been bounced off a couple of consumers perhaps, or bounced off internal people. I wanted to be able to bring forward quantitatively tested concepts of new product ideas that we had actually thought through on the technical side and had clear approaches on how we would go about the whole thing technically. But I wanted to make sure we had the consumer heartbeat established and established to the point that we had some quantitative concept consumer test results on that

So, looking at that as kind of the end point of what we wanted to deliver to SLT 9 months after the project was complete, we backed that into experience and project management. You know I had an MS (Microsoft) project time schedule and everything set up for it. And so we kind of cracked it out and said alright for the first three months we are going to do nothing but do ideation and collaboration on those ideas; and the tool we had used, recognising this was a quite disparate team, disparate in a sense they were still a number of R&D functions all with a day job to do and this was kind of an add-on type activity for them to actually put time against.

The team decided to create a team name or brand for themselves and they came up with *Curious George*, the 1940's cartoon character. In the original stories, Curious George, a monkey, was brought from a zoo in Africa to live in 'a big city' in the US. His appeal for the ISF team is that because everything in his new American world was so novel that he kept asking questions; questions which were sometimes very profound. This was the spirit the US team wanted to inculcate in themselves; a restless curiosity and sense of joy in discovery.

Below (Figure 5.7) is the Microsoft Project planning schedule for the US, Curious George team. The plan breaks the project into discrete phases with key dates and milestones indicated. The team leader developed this plan at the very beginning of the project; he shared it with the team members and they agreed to try to work to it so they would stay on track to meet the deadline at the end of the project.

Figure 5.7: The US Team Project Plan for the ISF Project

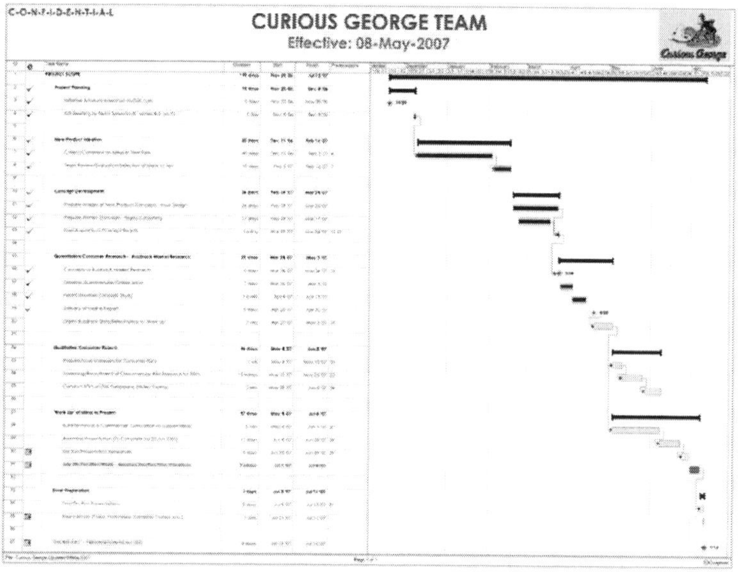

Source: US Team Documents.

Coapman arranged with the site's Facilities Management team to have a dedicated meeting room for his team and this project. Each of the team members got their own key. The team held their meetings in this room and eventually it was to become the storeroom for all their prototypes and competitor samples. From the first week, Coapman set up weekly meetings; two per week Mondays and Fridays, for two hours each. He wanted all the team's meetings to be face-to-face and this was possible given that all of the team members were based on the same site.

5.5.1 Getting Started

One of the more senior members in terms of seniority, Dr Buch recalls:

We had regular team meetings where we all sat in a conference room; one of the things that we did that I think worked very well is we reserved a team room and we had a dedicated team room where we could keep all of our materials and we met there weekly. In fact, in the beginning we met twice a week.

Coapman also acquired software licences for all of his team for a tool called 'ThinkTank' created by a company called Group Systems. ThinkTank is described as 'group intelligence software' and it allows people post up ideas via computer, generally in a very raw fashion regardless of where they are or the time of day. The ideas can be seen and commented upon by other members of that project and, ultimately, they can be voted on by the team members to determine which ideas have most appeal. Coapman describes the starting phase:

What we agreed to do is meet twice a week for two hours, we set up Group Systems as the key tool for doing the ideation and used that as a repository for people to enter, essentially cores of ideas, so it wasn't fully flushed out concepts, but it was people that had stumbled across interesting technologies, whether it be from scanning the store shelves at Walmart to finding infomercials at night to little blurbs on the internet, a whole variety of sources. Iconoculture was another rich source of some neat ideas that were coming out.

Just post into this ThinkTank session, the core of the idea and it was then made immediately visible to the rest of the team and they could easily just click on that idea and pose questions or challenges or improvements to how that might go forward. So we did that for about three months.

Coapman also tried to stimulate the team's thinking by giving them a voice recorder (dictaphone) so that if they were struck with an idea at any time of the day or night, they could record it. He also asked them to go out into retail environments that were not on their normal shopping itinerary and to look for new and stimulating ideas and bring them back to the group's team meetings:

One of the first exercises that I gave the folks to do after we kicked it off in, I guess it was January was, I had a grab bag and in this grab bag I had a piece of paper and one said 'Circuit City', one said the 'Dollar store', one said 'Wal-Mart', one said 'CVS', one said oh a couple of other things, it was a grab bag and people randomly took back and their assignment was to go out to that particular venue, look in an area that's outside of the standard health care arena and come back with at least an idea as to how you might develop a product that kind of stems from that idea.

Dr Chaturvedula recalls:

I think we used many tools – Scott gave us an audio recorder so we could record our ideas wherever we were. He asked us to select some stores and go and find interesting things – I went to Sharper Image and we started to look at different ideas. In the first stage, we were just throwing out ideas, nobody was judging – we were all just adding and building. We used Think Tank an electronic tool.

In this phase, the team devoted themselves exclusively to generating ideas. Dr Adusmilli recalls:

First step was coming up with ideas and we used a software package called Group Systems – which facilitates the building and adding of ideas.

Similarly, Dr Buch recalls how they, initially at least, tried to favour bold ideas:

Initially, what we did was we all just sat around the table and tossed ideas out and, you know, kind of wild and wacky stuff, the crazier the better.

Dr Schwartz also recalls how Coapman managed these meetings:

Scott created a high comfort level – people were very open, transparent, nobody was uncomfortable. He was very neutral – he was very good in terms of the psychology – people felt comfortable.

This theme was echoed by Dr Chaturvedula;

The team was very friendly, the environment was encouraging – nobody was turning off any ideas everyone was building on ideas.

Over the first three months, through the combination of face to face meetings and the online facility to post ideas remotely, the US team got up to 300 ideas. A couple among the group, somewhat cynically felt that this high level of ideas had been reached because most people came into the project with a number of ideas already. Dr Buch noted:

Most people seemed to gravitate towards and source their ideas from the business areas and therapy areas with which they were already most familiar.

Dr Lilan Chen agreed that the composition of the 300 ideas generated by the team included a substantial number that people had been nurturing for some time, certainly since before the ISF initiative:

> Generating the idea is pretty good actually. Basically many of us just came in with old ideas so that's why we would generate over 300 ideas so quickly.

Dr Chen also had another perspective on GSK's appetite for radical innovation that tempered her interest in looking at radical. 'blue-sky' ideas. She believed, from her experience, that the company would not support anything except incremental ideas and so she confined her own contributions to safe, near-term, line-extension type of ideas:

Although we were asked to try to come up with ideas or concepts which were totally radical different from the role we are in, the business we are in. And spend a lot of time and effort in blue-sky areas - I just didn't do it. I didn't think that the company would go for these ideas; I didn't think it would work so I did not support for those concept and ideas items.

Coapman observed that at this stage sometimes the high level of technical knowledge possessed by most of the individuals on his team was actually limiting the ideas that they were developing. They tended to curtail their thinking just because they knew what might be technically feasible. He found it refreshing to have one or two people on the team who were technically naive:

I think in the generation of the ideas themselves... the less versed people are in the technical side of things may have been a stronger asset for us, to get someone closer to just your general everyday consumer like Greg Smith (Regulatory Affairs) for example. R&D people tend to over think things and may not attempt ideas because they may know a lot about the technical limitations of things, it can almost sometimes can be a self-confining thing. ..You get someone saying well that is not possible because this, and that. So sometimes that was a bit stifling and it took some management to say, well put that aside, the idea unto itself, the benefits that it can deliver. Maybe the technology that is being suggested to deliver that don't; appear feasible, but there are many ways to skin a cat, so let's just look at the benefit that it delivers, the insight that could bring the consumer into the product idea, how you could talk about it, how you actually technically execute against it, let's park that aside for now.

5.5.2 Incubating and Prioritising the Ideas

At the end of the nine weeks dedicated to generating ideas; the ideation phase, Coapman directed his team, in accordance with their project plan, to stop having raw ideas and he decided to move them into the second phase:

> At the end of nine weeks we kind of said alright, we need to stop this now, we can't go on in perpetuity for ever generating new ideas, but you know in order to meet the timeline of doing some qualitative research on some of the promising ideas and the quantitative buzz back research, we need to stop now. So we essentially terminated the ideation phase and of the 300+ ideas that we had generated.

The team had three hundred early-stage ideas but they knew that the ideal final result would require them to reduce that to just 8 ideas, so they felt that they needed to begin a process which would whittle overall number of ideas down. They used ThinkTank again which allows participants to rank ideas and vote on them. While, for some ideas there was no market data, Coapman felt that the ThinkTank process was providing a layer of objectivity for these decisions. The team would individually rate the chances of technical success for various ideas and compare that against the possible market value of the idea and make their voting decisions accordingly. Coapman adds:

> We also used Group Systems as a way for everybody to vote in on, alright based on a variety of criteria; what is the technical feasibility, what do we think the commercial value is and that really was kind of a finger in the air type thing. Do we think it's $40 million incremental, do we think it's less than that. We didn't have any strong market research to really back that up, so it was largely, you know, intuitive of what people thought.

The group worked together using the electronic tool and through a process of discussion and debate, they were able to reduce the number of ideas from 300 down to 200.

Scott's prior experience in marketing both for P&G and GSK equipped him well to orchestrate the next phase of the research process. He employed an external company called Buzzback®. Their approach is to get an online panel whose demographics match those of the target audience for each project. They expose the panel to two ideas at a time and ask them to indicate which is their preferred one. They then use an algorithm to establish, after multiple exposures, a ranking as defined by the consumers. Thus, if you have a significant number of ideas and want to establish which ones have most appeal with your target audience Buzzback® offers a suite of tools to help. Traditionally, Buzzback is used if you have approximately 100 ideas and want to find out what are the top 10. But in order for ideas to undergo the Buzzback methodology, all the ideas have to be written and rendered (illustrated) in the same high presentational standard. They are referred to a 'concepts' and this denotes a clear description of exactly what the idea is; what benefits it offers, how or if it has been tested (i.e. 'clinically proven') and this information is often accompanied by an indicative selling price.

Dr Adusmilli recalls this element of the process:

We took 200 ideas into an online quantitative tool called Buzzback. We hired an (graphic) artist and a copywriter to render them as professionally as we could. We also fixed on a selling price for each idea. Each idea was seen monadically by 100 people – then we prioritised the top ones.

5.5.3 Researching the Value of the Ideas

Coapman understood, from his prior roles, that in (new product development) concept research, there are very established formats for presenting the ideas. The leading research companies have been carrying out research on new products for several decades now using similar methodology. Converting the raw idea, which can often be a technical one, into a crisp concept that a consumer can understand, is a skill that commands a premium in the marketing world. If done badly, a good concept can end up being killed purely by mediocre execution.

Coapman was very conscious of this and so he spent a high portion of his budget on specialist consultants who were on the roster for the (GSK) commercial organisation and who were expert in the art of concept writing. He was trying to successfully anticipate possible methodological objections that might be raised by the commercial team when they were presented with the US Team's ideas. Coapman was keen that the organisation would adopt his team's ideas and assimilate them into its pipeline. Hence, he wanted to present the ideas in a format which was congruent with how GSK manages its NPD ideas. He didn't want to lose the contest on the basis that he didn't use exactly the right testing methodology or the right illustrator to develop the concept. As he explains it:

One of the concerns we had with this whole project from the beginning was you know, it would be terrible if we came out with a bunch of great ideas and we tested them in a way; and then we brought them forth and we were told that it wasn't valid research because we didn't do it the way that the commercial folks would have done it; or we didn't test it with the right types of consumers or such. So we wanted to put our best foot forward and make sure, minimise the chances that the data we generated would be second-guessed and pooh-poohed so to speak. That's why we had intentionally selected people that the US business had already done work with, they had confidence in them, they'd proven themselves.

The Buzzback process allowed the team to get down from the 200 remaining ideas to a final brace of seventy-five ideas. With seventy-five ideas remaining, Coapman brought in a team, comprised of two professional illustrators to work with his team so that they could accurately capture the visual essence of the ideas being proposed. He brought his own team together for one of their regular meetings and they briefed the commercial illustrators. Coapman describes the process:

> We actually hired a group by the name of Insync Design which was conveniently located right up the street here. The Aquafresh (US Marketing) team had used this company before to do some visualisations for some of their Aquafresh things and so you know, it was a very easy person to engage in that. In a similar fashion, two of the illustrators from Insync Design came in and for two days sat down with us face-to-face; we explained through what the idea was, the concept, in our mind what the product would look like and they would be doing sketches right in front of us. You know rough sketches right in front of us and sketching them out and saying, is this what you mean and we'd say no, maybe a little different and they'd re-sketch it; and then they came back to us with electronic files in about two weeks time and we put the image to the concept, the articulation of the concept and made up concept boards.

Dr Buch recalls that Buzzback was helpful in eliminating some of the superfluous ideas that may have had a passionate sponsor within the group. This was an objective, scientific method of homing in on what consumers themselves felt to be the ideas with the most appeal.

And that is when we started to narrow the list down and we narrowed it down to about seventy-five and then really started focus in on those. We focused in from two aspects, one was the technical aspect – was it technically feasible and if so how would did we do it and then the second aspect – was to try get some consumer feedback so we started to think about consumer research for example, to get the consumer feedback on those seventy-five or so ideas. Then we narrowed it down from there.

He notes, however, that not everyone was prepared to agree with the Buzzback results:

We sought Buzzback's information input but we did not necessary agree with it all the time.

5.5.4 Difficulties Emerge in the Selection Process

Scott's team now began to experience some difficulties with this phase of their process. Despite having a process for selecting the most attractive ideas, there was still some friction around how the process was used and to what extent the results of the (Buzzback) concept research should be taken as the definitive guide to which ideas should go forward. The Buzzback process relies on an online evaluation of an idea. The respondent gives a very fast reaction; an almost visceral, gut-feel. These types of tests can favour ideas that are straightforward, already familiar and conventional. Dr Buch, who had been involved in this type of concept research before recognised this possible shortcoming of the process:

We realised that there could be some ideas that were just so far out there that Buzzback probably did not really probe it as effectively as we would have liked. There were several ideas actually that did not necessarily do real well on Buzzback that we ended promoting towards the top of our list anyway. That is just based I think on consensus of the group, we had a group fairly highly trained people who had been around the block a few times and I think that there was strong gut feel that an idea was a good one it went forward regardless of what the data showed at the early stage.

The group used their own judgment to promote some ideas that did not score well in research. They went on instinct and experience when it came to certain ideas. But Dr Chen found that this fluidity and reliance on personal judgment allowed certain people to preserve pet ideas and a number of ideas that she felt had been already exited, suddenly resurfaced:

We were going to screen each idea and somehow we didn't do it in a systematic way and I think that suited some members having their personal objective. What I mean by personal objective is, they probably send in that idea in the first place. They just want to carry out, they just want to carry and run with it. And then were used, sometime the medium wasn't managed very well. It got out of hand or discussion about someone propose their own, I mean just try to elaborate and sell their idea to a team. And they were going on and on for hours and hours and he or she not giving it up at all.

Dr Schwartz was also surprised that some of the ideas that she felt lacked the scientific credentials for the contest were allowed stay in. She felt that Coapman was not refusing any ideas, he didn't see it as his role to screen out any ideas; he wanted the consumers to have the ultimate decision:

Scott felt we shouldn't disqualify too many ideas – he wanted the consumers to decide. He didn't like to take the responsibility to eliminate ideas no matter how bad they were. Some ideas had zero science and they made no sense to me.

Dr Chen felt strongly that Coapman 'was too nice to say no to anyone' and that he ought to have shown stronger leadership in managing the exit of some of the ideas that she felt had been rejected in research. Dr Mishra (Medical Director, and one of the two most senior people in the team) agrees that Coapman did not lead this phase well:

Ideas kept coming back and this was a lack of formal leadership from Scott in ensuring the exit of the ideas and he wasn't adequately forceful. Scott should have played good-cop, bad-cop.

Dr Schwartz noted that people became too attached to their own ideas and continued to try to progress them even though the team felt that these particular ideas had already been rejected:

Even after we had agreed to eliminate certain ideas from consideration – nevertheless, they kept coming back. People were unwilling to abandon their own ideas.

One of the team members, who was mentioned regularly as being the primary offender in this regard, has a very different view of how to interpret the rebounding of ideas. Dr Assumilli notes:

To have an effective team, you're bringing in people with creative abilities – and usually they're very passionate about what they do and not willing to let go ideas they care about. When you have passionate people promoting ideas – they are bound to come back again and again – and I think that's a healthy thing. I think we needed the passionate people.

5.5.5 Finding Time for Innovation
April/May 2007

During this period, some team members also began to experience considerable pressure in their 'day job' and were finding it difficult to carve out the time to participate in ISF. Coapman took an active stance in managing this for some people. Having established a pattern of two physical meetings per week, each lasting a minimum of two hours, Coapman did not expect a full turn out at each meeting. Dr Schwartz noticed a decline in the attendance when they were going through this difficult phase:

> *Time was a problem – rarely had many people at each meeting mostly only four or six per meeting. Especially when people were arguing about ideas that we all thought had been killed.*

The team spirit seemed to sustain them through any specific pinch points that threatened to take members away from the team. They managed to compensate or even to work with the members to help them through particularly busy periods so that they could continue to contribute to the team. Dr Mishra notes:

> *My view is that people just managed it by working over 100% anyway. No-one held back because the team felt the sense of one-ness.*

In some cases, though, it went further than that, the team managed together and helped each other out. Dr Buch recalls:

I was told just to just do what you need to do and get it done and that is really what we did. I think most of the people on the US side realised that they just had to juggle things and do it effectively. I did hear a few complaints from people saying their managers were kind of getting on them about it. When that happened the team just sort of coalesced to take on others peoples responsibilities to free them up for a week or two if we had to. It was really a well functioning team where I think everyone had a good sense of camaraderie. People stepped in when they were needed and sometimes people had to step out to get something done that was urgent. But others just filled in for them.

Coapman did find a way to get people to prioritise the ideas, both their own and the others. Having done the first phase of quant research through Buzzback, the team were still left with a number of ideas that scored well and a number that didn't score so highly but were still well-regarded by the group because of their originality. Coapman notes that the spectrum of ideas was still very wide:

... there were 75 concepts and they were all over the map as far as therapeutic categories. It was smoking control, it was nutritional, it was analgesics, it was respiratory care, it was weight control. And then there were these bizarre outside anti infected gum control ideas, a variety of different devices.

5.5.6 Setting Up Consumer Focus Groups

May 2007

To break the deadlock, Coapman hired a professional market researcher as a consultant to the project. Again, it was someone known and trusted by the US commercial business. The researcher took the Buzzback results for the seventy-five products that had been tested and she performed some further analysis. She looked at the concept scores and derived a formula which she referred to as 'need-gap analysis'. This involved looking at a number of variables within the concept score and working out a calculation that combined the attractiveness of the idea with the importance of the need the idea is intended to fulfil. Coapman recalls:

The way that Lauren had suggested the best way to do this given the number of different variations of categories that we were in, was to calculate a need gap score for each of the concepts, so it was better able to compare on an apples-to-apples basis the general strength of the idea with consumers. So essentially you know, the strength of their interest in the product and the benefit that it provided was essentially a mathematical product of those two values. So at the top box for those two, we multiplied them together and got a need gap score. And that's largely how we prioritise which are the ideas that were the strongest. So those ones that were very strong, we went back and said now let's go back and throw these in front of consumers face-to-face.

This new algorithm got the number of top level ideas down from seventy-four to a more manageable seventeen.   Coapman then arranged for each team member to plan and moderate their own customer focus group.   This meant arranging a group of target customers to attend a research group.   Each team member would have to moderate the group and show their ideas, using the opportunity to get customer feedback on their ideas.   Coapman thought this would be a useful learning opportunity for his team of scientists to get in front of consumers and to show them their ideas. He was confident that the consumers would automatically push back on ideas that didn't appeal to them and this would be a good mechanism to cull some of the poorer ideas and allow the team concentrate on the ones that emerged as the strongest from this process. The idea of standing in front of customers was a very new proposition for almost all the team members whose natural R&D home is generally in the laboratory.

They were each given some training from the professional researcher so they knew how to handle the consumers in the group and Lauren, the researcher, managed all the logistics around recruitment, scheduling and the provision of the right stimulus material.   But the scientists in the group still had to run the group and manage the dynamics of a focus group.   The research groups were videotaped. Coapman recalls:

> Few, if any other than myself, had ever either attended a focus group as a viewer, you know most people had maybe attended a focus group but some had never attended a focus group, not to mention actually being out in the front room with consumers and talking with them, people found that to be extremely valuable, so much more beneficial to their understanding of the idea of when they are talking with them human being to human being, person to person and getting the scoop right from the horses' mouth.

One of the team members, a medic from the Regulatory Affairs department was very daunted by the prospect of running his consumer group.   However, he proved so good at it that the organisation recognised this talent subsequently and have used him to do focus group research outside this project. Coapman recalls:

> I think some of the groups that Greg Smith actually facilitated were fantastic, he has got kind of a very natural speaking ability to connect with consumers. I think that was a hidden talent that he never was aware of and he has since been recruited by the medical affairs group who are doing some focus groups and they actually recruited him to actually do some Breathe Right focus groups. So I mean he really enjoyed doing that, I think he did a fine job of it and so I think that was a hidden talent that he never knew existed.

The others also found value in presiding over the focus groups and exposing their ideas to potential customers. Dr Mishra remembers this aspect:

> The experience of talking to consumers was very uplifting, energising, very motivating, very, very - there was an end goal in mind and that was very powerful.

Dr Buch had some experience of attending focus groups before but he enjoyed seeing the others experience this element of consumer interaction for the first time.

> It was not quite as foreign to me I had done lot of that, but it was really fun to watch some of the people on the team who had never done it. Because you could see the light bulb come on. It was truly an epiphany. As an observer of them I think I could see it in them almost more than some of them realised it in themselves.

Dr Chen was one of those who hadn't done this before and she explains how she learnt a lot from it:

> Oh yes, yes do you know I conduct a focus group that's very interesting to me. Yes before I had only some vague ideas about the consumer and how

research is done but after the ISF I have more idea of what concept testing really is and how to test this and then how to read the test result ok. And then how to put the whole thing together and then get the consumer research knowing about how the consumer feel, how they respond to your concept. I'm a technical person and I'm able to come up with a modified concept immediately I hear the feedback and then, ask their input again and they respond again. So it was very helpful for me to actually tell me to actually design what will be the actual product alright. And yes I have more thorough understand about the how the idea being initiated and been tested and being refined and how can I also offer our designs for an actual product to meet consumer's needs. Yes I do learn a lot from the focus group experience.

The focus group process did provide the basis for the elimination of certain ideas and the promotion of others. These sessions were videotaped and played back for the other team members and this seemed to ease the passage to consensus that had been so difficult to arrive at prior to that. Dr Assumilli also recalls how the consumer research helped the team to flesh out their ideas:

I enjoyed the consumer interaction so that when we presented them, there was a lot of meat on the bone in terms of consumer research. Without the consumer piece, senior managers could not make good decisions on which ideas to progress.

June 2007

Now, with the slimmed-down list, Coapman decided to engage with some of the internal stakeholders who would ultimately be evaluating these ideas. He decided to informally approach the VP's of R&D on the key business Future Groups to make sure that the ideas his team were going to propose were not ones which had been explored before. He explains:

We did some balance checks because ... we really didn't have visibility as to whether or not these were truly new ideas or whether they were something that they (the Future Groups) had looked at before. You know there is no huge central repository of every idea that has been looked at across the organisation, so some of these ideas may easily have been repeats of things that may have been considered. So rather than spin our wheels and go back reinventing the wheel... we did convene with the commercial and R&D futures heads... and said here are some ideas that are being batted around, we think there is some value to them for these reasons, would they be on strategy, would they be off base from what you consider? ...just to make sure we weren't reinventing the wheel and going down a path that they already had data to say this is a non-starter for us.

5.5.7 Preparing for the Presentation Day

The group now started to focus on the remaining ideas and how to express them. They started to think about the presentation itself and how to prepare for it. They kept the graphic designers and the copywriter on board and really refined their key concepts. In the end, although the brief suggested eight ideas, the team presented seven. Coapman managed the choreography for the final presentation:

We figured there was about seventeen of the seventy-four that performed quite strongly that we recommended we should take forward. Now the reality was that when we presented to the SLT we had I guess four hours of time and there was no way to do justice to those ideas in appropriate depth in that period of time, so we culled it down and presented ... seven to the SLT, but reminded them and say there's about a dozen or so more that we feel very strongly about. And after that presentation at SLT we got in some further details and briefed the other VPs on these are some additional ideas that we didn't flush out when we met with you in the formal meeting, but we think they are strong, strong information there to go forward on.

Coapman and his team chose the seven ideas to formally present to the SLT at the final presentation. Their choice reflected a balance between trying to get one idea in from everyone on the team while also choosing the top performing ones overall. The remaining concepts formed the appendix in their presentation slide deck and document. The US timeline is represented below (Figure 5.7). It tracks the number of ideas that were generated and then follows the process as they advance towards the final presentation.

Figure 5.8: US Idea Management Process

November 2006   July 2007

ThinkTank   Need Gap Analysis   Group Decision

300+ Ideas   200 Ideas   75 Ideas   17 Ideas   17 Ideas   7 Ideas

Initial Ideas   BuzzBack   Focus Groups

US ISF Idea Management Process

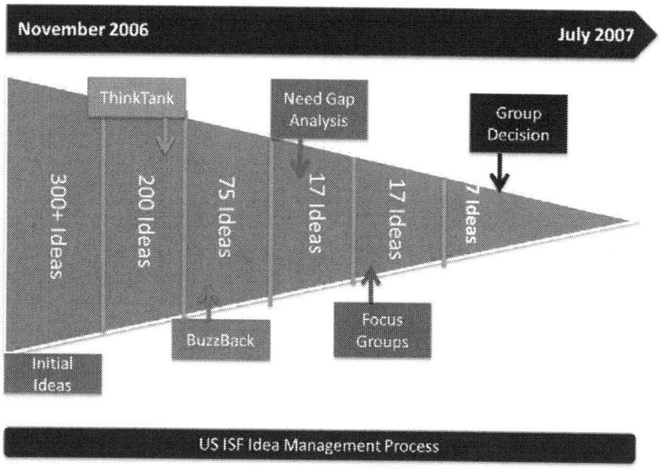

Source: Case study documents

The final presentation took place in Parsippany in a conference room within the company's major R&D site. The date was the 7th of July, 2007. A full day had been allocated to reviewing the presentations. Both teams were present along with the SLT and some of the company's senior commercial team. The UK team presented in the morning, there was a break for lunch after which the US team's ideas were presented. At the end of both teams' presentations, the SLT congratulated both teams and provide structured feedback on the ideas. There was no official 'winner' declared at the meeting and, indeed, within the SLT opinions were divided about which of the teams had done the best job.

5.6   The Big Day

5.6.1 How the Teams Rated each Others' Performance

Most of the R&D people would have had some connection with each other through various project teams over their time with GSK. However, as there had deliberately been no contact between the teams for the duration of this initiative, they were most curious to see what the other team had come up with. Those that did travel from UK for the presentation found it fascinating to see how the US team had gone about it and just what they had come up with. In the teams' introduction to their presentations, they revealed their ways of working and it became apparent the amount of time each team had devoted to both generating the ideas (ideation) and evaluating them. Dr Uruquart (UK) recalls:

For me Nigel's process actually to some extent overdid the ideate and under did the evaluate. And I actually think when I looked at both processes I'm envious of some of the things that the US got to do but I would not have wanted to be in one of their groups.

I guess what I am getting to is I love the ideate rather than the evaluate. I can do the evaluate. I would have liked to see more evaluation built into ours, but I wouldn't have liked it to have gone to the point where Scott went to where frankly your squeezing ideation band to a couple of weeks in order to get everything evaluated. And there's a happy medium and I don't think either of us actually found it.

Grist also recognised that the more structured approach of the US team might have some advantages:

We knew they were going to take a slightly more pragmatic, doable approach and we took a rather more ambitious and adventurous approach. I have to say that I was impressed with what they did and I was pleased that what the two groups had come up with and done was different.

The obvious overarching structure of the US approach was the first element of the presentation to make an impact on the UK team. It is the process and the not the ideas that seems to strike the UK team first. Dr Stovell recalls his reaction:

There was quite a difference between the UK and the US teams approached the whole thing. I mean the US had a much more structured approach, they had dates by which they had to stop having ideas for example, whereas we were effectively having ideas right up to a couple of days before and some stuff made it into our final presentation that had only come to the table – well in fact there was some stuff that went into the presentation that none of us had actually seen other than the person who presented it. So our process was wide open.

Dr Uruquart also noted the process as the principal thing that separated the two teams. He envied the US team in some of the items they had managed to achieve, primarily the consumer research. But, overall, he felt that he, personally, was more comfortable in the idea generation phase than in the more analytical phase:

Me, I couldn't have coped with Scott's system, I would have gone mad. I mean, because for me the whole beauty of this process was that we were free to go and explore and Scott's process by going so 'end-to-end' basically totally minimised the amount of time they could explore in order to maximise the times that they could evaluate. For me Nigel's process actually to some extent over did the ideate and under did the evaluate. And I actually think when I looked at both processes I'm envious of some of the things that the US go to do but I would not have wanted to be in one of their groups.

I guess what I am getting to is I love the ideate rather than the evaluate.

Grist could see the differences in the ideas as well as in the approach. He was pleased that the quality of his team's ideas seemed to stand up to the competition pretty well.

I was really pleased actually when I saw their ideas because they were so different. I When I saw the presentations I was actually really pleased that they were so different, because that was part of the experiment and I thought there were aspects of what they'd done that were better than what we'd done, and I looked at what we'd done and thought there were aspects of what we'd done that were more advantageous than theirs.

So I was just really pleased that we hadn't all trotted out the same stuff, because I thought that would have been disappointing. We knew they were going to take a slightly more pragmatic, doable approach and we took a rather more ambitious and adventurous approach.

Dr Jon Creeth remembers his impression of the US presentation:

The contrast in how the teams went about it was amazing. One, ours, was very emotional, full of visual appeal and big ideas and the other was very logical, linear and methodical and a bit predictable.

Dr Smith reflected after the US presentation that he felt their ideas were weaker than the UK ones and he attributed his perceived difference in quality to the styles of leadership of the respective leaders:

Our ideas were more creative although I suspect we could have had a fraction more process, but overall it didn't matter too much. I think it was better to have less process and keep the free flowing, the chaotic bit of it going, than have more process and stifle the creativity.

In the US appraisal of what the UK had developed, Dr Buch (US) recalls:

It was really interesting when we went the final presentation to see the difference between the UK approach and the US approach. I think the US approach appeared to be a bit more structured. I don't know which one actually was correct, I think the UK approach demonstrated a tremendous amount of creativity but I think the way they approached their presentations weren't quite as defined as ours. On this side of the pond I think we went into a bit more of how things could be executed. Whereas the UK team went into more of the pie in the sky really unique idea type stuff.

Coapman also was slightly surprised at the way the UK ideas were presented. They had not followed the format he had expected:

The UK team they took the approach, and not right or wrong it's just they took a different approach, and they kind of looked at some more global areas of interest that GSK might be considered in. So they didn't come back specifically with defined product ideas, concepts validated, you know? For example, sleep was an area that they thought could be a big opportunity here for GSK, sleep disorders. Should that be a therapeutic area we should be getting into? People just don't sleep as well as they should today, they don't sleep as long, they don't sleep as well. Lack of sleep interferes with productivity, lack of sleep interferes with energy levels the next day, so they took a kind of more holistic approach as far as what are some of the therapeutic areas we should be getting into and then there were some examples of some ideas that they thought would help fill that space.

Dr Chen was also impressed with the creativity of the ideas of the UK team but was surprised that they appeared to have failed to answer the brief, specifically the requirement to have some estimate of consumer appeal and market size:

To me, I was very impressed by the UK team. UK present more concept and idea ok but they don't hit the science a little bit ok and they hit the – I don't think they count down any consumer research ok. ... Ok but then present many great ideas actually. I was very impressed with the concept and the ideas that they have come out. The quality, the numbers and then the area which is the idea then get into it. I was very impressed in the way they presented it too. But given they were told that the way that the SLT wanted them to do it or either the way they want the favourable objective which is the...we were briefed to present need to have some scientific rationale and then estimate the possible market consumer needs and then market opportunity and then what will be the possible time for launch, something like that. They did not meet those criteria which is true SLT did give us some guideline about.

5.6.2 The Judges' Decision

July 2007

The presentations were made to the R&D SLT with all the key members of the team present to hear and discuss the ideas. Each team was given the same time to make their presentation with the UK choosing to go in the morning and the US taking the afternoon. The UK team had prepared some visual stimulus for the room with cardboard cut-outs of some of their ideas. They had also assembled a collage of posters and images of science in technology and in nature. They had posters made of these and these were hung around the room.

The US team used a lot of video footage of their team talking to consumers during their presentation. They had less contextual imagery and more project-specific content. The SLT paid careful attention to everything they heard and saw and the casting vote on which team had delivered on brief better rested with the Senior VP, Stan Lech. He sided with the US team and concluded that they were the competition winners. However, this was not formally announced and there was no winner or loser declared on the day. The role of SLT on the day was to provide a positive reception for the ideas and to discuss which of them could be brought forward to implementation and how it might be done.

Afterwards Lech explained why he favoured the US submission:

Honestly the edge went to Scott's team. I think it is because it was structured and they had data and they had consumer research, they had pretty much concrete data from the consumer. They ran it like you should run a project. Scott's structured approach was very much on brief. It was like this is what we can do for the future groups that we think we have a 25 million and these are some opportunities that were new businesses.

The UK just looked at pretty much new businesses – again I think because Scott was very structured and he took the brief as we told him and the UK said, no, we're going to go after the big ideas, we're not going to spend time on the Future Group. And that was just the way it unfolded. And I can also tell you that the workflow from the structured versus the unstructured was very interesting too. Scott was very much very methodical about time, its meetings and scheduling and the UK did like the cram. Like the college cram. They would get together and just brainstorm intensely come back and them cram again for hours and hours and come back where Scott was kind of metered. So there was a very different environment set-up.

Lech is reflecting that the UK team did not deliver on the brief which he had thought was specific and unambiguous. They did not generate ideas for the five big brand groups; the Future Group. Instead, they focussed on entirely new ideas in entirely new categories. He notes that they were considerably less structured than the US team and he equates this with their failure to adhere to the brief but he does acknowledge that they (UK) did bring a lot of creativity to the table:

> *The people in the UK just tend to be a lot more free spirited and less structured and that unstructured approach created a lot of conflict I think. It's a very interesting thing in the UK they very rarely got face to face. – Why? Because they could not arrange their calendars to get face to face. But if they would have forced it they would have said – look this is a very important thing you need to do folks, we have got to clear our calendars. They would have forced things to fit, right and then they would have had other things not happen. Which I think is what the US did; I think the US forced ISF to be a priority. As well as having the structure which Scott provided. The end of the day they both did really great. It's just when you play it back the US was more structured and it just tended to come over a bit more credible than what the UK presented but the UK was very high in creativity.*

Lech acknowledged that there were, as he puts it, 'stronger personalities' within the UK group and these may have been harder to manage than the people on the US team.

> *There were a lot more rebels in the UK and the people in the UK were a bit more out of control, and that's kind of reflective of their personality behaviour. For Nigel, it was like herding cats....But there's some big personalities in the UK. If I look at the personality metre, there's no question that personalities in the UK were much stronger.*

Stan's own direct reports did not share his opinion that the US had won the contest. Many of them, in fact, thought the opposite. Simon Gunson, VP of New Product Research reflects:

The UK team were wild, wacky and weren't inhibited - but the US team, driven by that structured approach – it was all facts and data. It seemed very much driven by potential science opportunities. The UK team were prepared to go that little bit further. A lot of the US stuff had been done before and a lot of the UK stuff was completely off the wall, in a good way, new – just what we wanted.

Another of Stan's SLT, VP for Panadol R&D, based in US was very critical of the US approach. He made the point that the group had been deliberately comprised exclusively of scientists. Additionally, they had the generation of scientific ideas as their end goal and he was disappointed that they didn't focus on the science instead of spending so much time working with consumers.

What the US did was put an awful lot of effort into making videos of consumers. What I wanted was some rigour and depth of work on the basic science.

He was most animated about this and was not the only one to hold this view among Stan's senior team, he continued:

What I saw was that a lot of effort had been directed towards consumers – what I was looking for was far more of the scientists and far less of scientists trying to run focus groups. I didn't find this refreshing because a lot of money and a lot of effort went into this ISF project. I didn't think this was appropriate because this was not a training exercise. As scientists, they ought to have focussed exclusively on the science and not got bogged down in marketing.

Simon Gunson (VP for NPR) agrees with the opinion above and is in opposition with the US approach of spending so much of their time and budget with consumer research:

I prefer the Nigel (UK) approach – if we look at innovation in its truest form, some chaos is necessary. I absolutely believe - that if we had a commercial team – then, yes, research was appropriate but there's no point in making first rate scientists become second rate market researchers.

Go out there, innovate, find the scientific opportunities – if we like them – then, fine – we'll go off and do the market research. But in my view, that's not what R&D are there for. They're there to develop scientific ideas.' I think there should have been far more effort on the science and only after that has been developed, evolved should you engage with consumers. But they spent effort engaging with consumers despite the fact that this could have been done better by professionals. But the ideas should have been well developed first.

I was expecting more, better science from scientists. They diluted the science by engaging with consumers. We would have judged the science.

Reflecting on the UK approach, Dr Layer, of the SLT, expressed surprise that although the teams had been explicitly been exempted from having to report their progress, she still felt that they (the UK team in particlar) ought to have checked back with her or one of the other SLT members:

It was all very underground when it was going on. It appeared to be quite underground and I had thought even though our brief was, you were not to step in and manage this let them be. I still thought, I had expected that people would give me a shout and just tell me where they're at and no one did. And I have to say it was an element of surprise for me.

Dr Layer also criticised the lack of project prioritisation from the UK team in particular. They showed a wide variety of potentially valuable ideas but did not place them in any apparent rank order:

Yes I do, I do think we got value. Do I think we'd have had better value if things had been better structure? Yes I do, but as I think we had to sift through and some of the ideas were wonderful, they really were. But at the end of the day it came back to we had to decisions on prioritisation. At the end of the day we didn't have a modus operandi of being able to progress all of the ideas. We had to sift and sort and make selection decisions but we didn't necessarily have all the facts and data but I would have presented back if I'd been on the team a rank order of their ideas. I would have said, my top three are X, Y and Z. But what we got was quite a long list of stuff. And I think what would have been excellent to have had would have been a list of the ideas and then share them all. And then say: And here's what we believe are the, if you're only going to pick two to go with it's these two.

In conclusion of this element of the narrative, the final word should go to Stan Lech, chief architect of the project. Lech asserts that the project was a significant success for the business as a whole. He confirms that the company have managed to get seven big (circa $20m) ideas out of the ISF project. He also cites it as a success on another level. Apart from delivering potentially lucrative ideas, the project also gave him visibility of the various talents within his department.

It was phenomenal. We had at least, well, let me tell you. We had fifty-seven ideas, we chopped thirty right out of the box, we then took forward seven ideas*, which are currently baked into either new categories, strategic growth initiatives, or a future group. So, I think we ended up with seven of the ideas of high interest out of fifty something, and three of those are going to be actively pursued as new category ideas, and I believe four of them are already baked into the future group kind of exploratory plan. So, seven more ideas, I think that's pretty damned good!

Also, I have better visibility now of the real strengths of some of these people too. So, it was good for us to be able to assess talent. I've had a discussion with at least two or three people on ISF about their future in innovation, because I was so impressed by what they had done that I can see that they have a bigger package than what I thought they had.

*Four ideas from the UK and three from the US team.

Chapter Six
Themes and Analysis

6.1   Introduction and Purpose of this Chapter

'There are many reasons why attempts to "manage" or facilitate innovation are fraught with difficulty' (Storey and Salaman, 2005. p. 31). The narrative of the ISF experience brings to the fore a number of contextual factors connected with this initiative of which some were enablers and others blockers to the successful practice of innovation and, specifically, new product development. These contextual factors provide a vivid and practice-based illustration of why innovation and, specifically, new product development, is widely considered to be so difficult to perform in a way that is predictable, repeatable and systematic within an organisation.

The purpose of this chapter is to identify the issues that emerged as most significant within the case study. It will describe these issues in the actual words of the team members. By identifying and classifying the individual themes drawn from the case study, this chapter will assist in categorising what are the key practical, managerial issues which are likely to either facilitate or inhibit innovation and successful new product development in large organisations.

This case data provides insight into the practice of new product development which might be helpful in devising better guidelines for future managerial practice in innovation in large organisations. The research question centres on how innovation actually happens inside large, global, R&D intensive organisations with a particular focus on radical innovation; how ideas are generated, incubated and implemented and how teams can best organise for innovation. This analysis of the case data provides insights into the questions

268

that informed this study, including: how firms and teams prepare for innovation; how managers develop practices and processes which are helpful to firm-level or project-level innovation and how leadership, in such a context, is operationalised and how firms can create team or project-level infrastructure conducive to innovation.

Given that innovation, specifically new product development, is a complex and non-routine activity within organisations, it is not surprising that there was a wide range of significant themes that emerged from the fieldwork interviews. Some of these issues seem very prosaic, intuitive and straightforward and would not, at first glance, appear to warrant a place on an inventory of innovation management practices. Yet, they were reported as having a significant impact on the performance of the teams and hence on the outcome of the overall project. Indeed, it is an advantage of this type of case-based research that it allows practical considerations to come to the surface which might be less likely to be detected using broad brush quantitative surveys unless they were specifically probed.

The issues raised can be grouped or clustered into four discrete classifications (see Table 6.1). First (Section 6.2), there are the issues of team set-up and initiating structures. Second (Section 6.3), issues emerge which come under the heading of processes, and, these include issues like meeting frequency and location, the use of IT and the process for harnessing the voice of the consumer in the project. A third (Section 6.4) significant issue centres on networks both within and outside the organisation. The fourth (Section 6.5), and possibly most significant set of issues, involves leadership and the leadership styles that were applied to these teams at various stages during the project. In terms of sequence, this chapter will begin with the issues of initiating structures and move towards what transpire to be the overarching themes of networking and leadership.

Within each theme described, evidence from the case study will be presented: first, from the team member interviews; next, from the interviews with the senior leadership team as well as quotes, where appropriate, from the company's close-out report on the project.

Table 6.1: Classification of Case Study Themes

Themes	Constituent elements
Structure	Making time for innovation
	Squaring-off line managers
	Size of team
	Space to innovate
Process	Ways of working
	Harnessing the voice of the customer
	Objective methods for idea prioritisation
Networks	Bridging Ties (internal)
	Boundary Spanning (external)
	Recruiting innovation champions
Leadership	Channelling the team's work
	Structured V's loose management

6.2 Project and Team Structure

6.2.1 Devoting the Time to Innovation

No issue was raised with more frequency, in terms of the number of ISF interviewees who raised it, than time. While, it seems so elementary and rooted in common sense, the way that this issue, of limited time, was handled in the ISF initiative was such that, according to the participants, it had a considerable, negative impact on both the levels of participation and the quality of engagement of team members. The issue surfaced in a number of ways. First, the team members were asked to devote 20% of their time (or one day per week) to the project but no provision was made for the accommodation of their other, prior, work commitments. Their workload or objectives were not adjusted accordingly. In practice, this meant that the team members had to both deliver their original objectives as well as working on the ISF project. Some felt that this was not feasible; they couldn't juggle their diaries sufficiently to release the time required for ISF and reported that they were left with no alternative but to retire from the project.

Additionally, the way the project teams were formed and the members chosen, the Senior Leadership Team (SLT) omitted (what transpired to be) a critical step of advising each individual's line manager of both their appointment to the ISF team and of the strategic prominence of this project in the view of senior management in the company. As a consequence, the line managers were often very negatively disposed to the ISF project and were, in some cases, actively discouraging their direct reports from working on their ISF assignment. This meant, ironically, that the devotion of time to the ISF project, for some, became a subversive activity, which had to be conducted in an almost clandestine manner.

A third factor was that those in the organisation, and on the ISF team, who had line management responsibility for a high number of direct and indirect reports were especially constrained in the time they could devote to ISF without compromising the level of attention they were allocating to their supervisory duties. This gave rise to a situation where the more junior R&D people were able to engage more with ISF and devote more time and energy to it. Although some senior managers, including a Vice President, had been purposefully selected for the group, their contribution to the team became severely limited and many of the more senior members, especially in the UK group, had to abandon the ISF project entirely due to lack of available time. In practice, there was a rough and inverse correlation between the seniority of the member and the time they were able to commit to ISF. This was particularly the case with the UK members. There were two types of withdrawal or departure from the teams of which the latter type was far more common; either a formal declaration that the time commitment was too heavy and that the individual could no longer continue to participate in the ISF programme, or alternatively, a gradual but informal drift away from the team and the project. More members defected from the UK team than from the US one but there was no significant difference in the seniority between the teams.

Table 6.2, below, captures some of the comments from team members where they allude to the issue of time. No individual is quoted more than once. It is apparent that within the UK team, this was more of a critical issue than for the team members in the US. This may be because of the more structured approach adopted by the US team which gave high visibility to the level of member participation right from the start. The US team met and connected in person far more often and they were able to discuss availability issues and help each other out. Nevertheless, time was an issue for both teams. Moreover, in the company's project close-out report, this is acknowledged as a serious flaw in the project design. This flaw is acknowledged alongside the separate but related issue around not briefing line managers.

Table 6.2: Finding the Time for Innovation

Issue and Source	Direct Quotes
Time Availability UK Team	*"I withdrew from ISF very early - never really got into it - due to lack of space in my schedule."* *"Time was a key barrier – twenty percent of your time was just not feasible. Great innovation doesn't come easily out of thin air – it's a combination of great curiosity and a great deal of thinking. One day a week was the wrong way to start it off. You can't come up with a great idea just because today is 'innovation day'...."*
	"Time was a huge problem. I ... was thrown in the deep end in a new job and to do this as well: it required more time. It was just at the wrong level; the worst level – if it was more than 20% we could have got more involved – if it was less than 20% then you could fit it in."

Issue and Source	Direct Quotes
	"Biggest flaw was that we didn't have time to do it. The one big learning is that you allow people enough time or you don't do this at all. Many people couldn't commit to the meetings and didn't contribute as they might have." *"So frankly there were a number of people who dropped during the way because they couldn't commit any time to it, and there were a number of others that did actually go to the end, but frankly committed hardly any time to it whatsoever. People's allocation of time to the initiative was inversely proportional to their seniority in the organisation."* *'My key learning was that I didn't have the space to begin to do the kind of job I wanted to do and if I am asked to participate in a similar exercise I would negotiate very hard on what I give up doing whilst on the project, rather than having a vague offer "to look at it".'*

Issue and Source	Direct Quotes
Time Availability US Team	*"I did hear a few complaints from people saying their managers were kind of getting on them about it. When that happened the team just sort of coalesced to take on others peoples responsibilities to free them up for a week or two if we had to. "* *"Time was a problem – rarely had many people at each meeting mostly only four or six per meeting. '* *"My view is that people just managed it by working over one hundred percent anyway. Nobody held back because the team felt the sense of one-ness."*
Time Availability: SLT (Senior Leadership Team) Comments	*"In the UK, I felt the guys felt they had enough time and the people we had on it were prepared to put in the little extra needed."* *"Culturally, in the US, they need to be more prescriptive in the way they work – they need their manager to give permission – that's why in the US there was an overall much more structured approach than in the UK."* *"People's managers weren't told what was going on. It left some people isolated as their managers may not have known - so it was run like a secret society; some of their leaders, understandably, felt undermined."*

Issue and Source	Direct Quotes
Time Availability Close-out Report Comments	*"Majority of time given was own rather than 20% of job hours."* *"Communicated by line managers that it is an out of hours activity, lack of clarity between SLT and line managers."* *"Some managers repeatedly communicated to focus on day job and not ISF and some individuals were made to feel guilty if working on ISF"*

In summary, very many of the team members voiced their perception of not having sufficient time to devote to the project. Nevertheless, both teams managed to deliver a portfolio of ideas that were sufficiently impressive that the senior leadership team declared the overall project a success. The issue of not managing the reporting lines with the line managers was a deficiency in the project design. This was reported to have caused unnecessary stress for the people involved.

6.2.2 Size Matters in Innovation Teams

At the outset, both teams were formally comprised of twelve or thirteen members each. But in each case, it was a core group from within that number that did most of the work. This transpired mainly for two reasons. First, three members of the UK team were located at a separate site and all three eventually disengaged from the project, reporting that they were feeling somewhat disconnected from the main team. The fact that their team leader did not create any processes to channel the work of his team was undoubtedly a contributory factor in their feeling of isolation. Had the project required a more intense, full-time engagement from the group located off the main R&D campus, their gradual isolation from and ultimate abandonment of the project may not have occurred.

Second, many of the participants had a high number of direct reports and they reported finding it challenging to juggle their 'day job' with the time commitment required for ISF and they consequently elected to abandon the project. The combination of both of these factors prompted a number of the participants, in the interviews, to voice the view that a group comprising 6 to 8 would have been preferable to the larger group which had been originally assigned. A third element to emerge from the interviews was that with too many on the team, there was an opportunity for certain team members to 'coast' and contribute very little to the programme. Those who were so disposed could ride on the coat tails of the rest of the team without having to make much personal contribution because the commitment to the group was relatively loose and there was always the option of leaving the group if the workload got too demanding.

The issue of size of team is not easy to separate from issues around both the duration of the project and the question of whether people should have been seconded to it on a full time basis rather than for this notional twenty percent of their time. By itself, the team size, as a factor may not have been so problematic. But, when it was coupled with such a lengthy commitment to a project for a notional twenty percent of people's time, it certainly was reported as an issue for almost all participants. Within Chapter Five, it is apparent that the UK team began to build momentum and to perform to a much higher level only after they had formally requested the 'drifters' to leave the team altogether. Once the team had been reduced to eight from the original thirteen, their performance improved.

In essence, the question seems to boil down to whether it is better to have a (relatively) large number of people with partial commitment to an initiative for a long period or whether it's better to have more intense engagement with fewer people working full time and over a shorter period. In hindsight, many team members came to the view that it would have been better to have had the exclusive, full-time devotion of fewer people for a shorter time period.

This view was consistent with almost everyone in the UK team where the team was not all co-located. These sentiments are reflected in the direct quotes included in the table below (Table 6.3):

Table 6.3: Size of Team

Issue and Source	Direct Quote
Size of Team – UK Team	*"Twelve/thirteen is too cumbersome – somewhere around six or seven feels optimal."* *"Thirteen people is too many; you can be hidden away without doing enough and remaining anonymous. A couple of groups of 5 would have been more effective."* *"There were thirteen people nominated onto this team who were each supposed to be spending twenty percent of their time. I would argue very strongly that the ideal number of people for this type of activity is six; you might go with seven in case you lose someone."*
Size of Team – US Team	*"So our core group was around six-eight and it was about the right size actually. We all could sit around the conference table and look one another in the eye and everyone had an opportunity to speak. It was about right."* *"Most of the work was done by a core group. When you have too many it's hard to get decisions made and to get things done."*

	"The one change I would make would be to have smaller teams but to dedicate them full time to the project."

Issue and Source	Direct Quote
	"I really think one of the advantages is that my team had was that we were able to get together almost on a minute's notice and go to that room and brainstorm and that is impossible when you are located at a distance like Coleford (the UK second site involved in the project) is."
Size of Team - SLT	*"It's not about the size – it's more about the diversity of expertise. I would think it's more important to have the variety of specialities – medical, regulatory, R&D, Formulating, clinical, project-management."* *"First, on the time front, I would make sure these people felt that this wasn't an add-on to their day job. I would free them up entirely for this.*
Size of Team – Close out Report	*"Teams of eight are the ideal numbers" (sic)*

6.2.3 The Space to Innovate – Having the Right Environment

Within the transcripts from this case study, there is ample evidence to support the idea that physical environment was felt to have an impact on the management of innovation projects of this type. It seems that the fact that innovation, by its very nature, is non-routine and requires that those involved probe into new areas, suggests that this is harder to do while remaining in the same physical environment where the routine work is normally carried out. No doubt, there are both physical and psychological reasons for this. Physical reasons, in this particular case include, the logistical concerns of needing somewhere to store prototypes and competitor samples. Team members wanted somewhere they could meet and discuss ideas as well as show and demonstrate prototypes and technologies. The allocation of a dedicated room for the team and the project was the operational outcome in both cases in UK and US and this seemed to have considerable positive benefits. Apart merely from the physical facilities that the room offered, having a specific meeting and storage facility also seemed to enhance the team members sense of identity with the group and the project.

This case study also exposes some of the difficulties in managing innovation teams that are geographically dispersed. In this case, the UK team were located across two R&D sites, situated about 300kms apart. But although they were geographically dispersed, and that constituted a challenge in itself; they did not have either of the compounding issues of different time zones or different languages to contend with, which is increasingly the case for some large organisations. Three quarters of the team were on one site in Weybridge while just three members were in a separate site in Coleford. The larger group managed to secure a dedicated innovation room, on site, assigned to them and were able to use that as the project head-quarters. Individuals used that room to work in and they were able to hold what few meetings they had there. The Coleford group however, because they were only three in number, were not able to secure any dedicated facilities on their site. Additionally, possibly associated with the relatively 'laissez-faire' management style adopted in the UK group; the Coleford team gradually reported feeling isolated from their team resulting in them separating and ultimately abandoning the project entirely.

This outcome; the failure to harness the potential of the smaller group located off the main site, has implications for the leadership of innovation projects in GSK and perhaps in more general contexts too. It suggests that when creative work is expected of the team, a higher level of connectedness, whether physical or virtual is desireable. It suggests that those members, not located with the main team, should have an almost higher or more potent affinity with the project vision to sustain them through the project in the absence of the personal, face-to-face interaction with the other team members. In this instance, given that the entire team never actually met face-to-face, it seems that the UK leader's expectation that all the participants would rise to the challenge and deliver on what he expected of them was somewhat naïve, especially in relation to the participants who were not located with the main group (Table 6.4).

Table 6.4: Creating the right space to innovate

Issue and Source	Direct Quotes
The Space to Innovate UK Team	'We eventually got our own project, team room which was a great place for us to work and store work-in-progress materials.' With so many ideas, we had lots of prototypes and samples and these took up quite a bit of space so we needed somewhere they could all be easily accessed by all team members. 'My direct boss wasn't so keen on my involvement, so it was great to have a dedicated room to disappear to. I could retreat in there and do my ISF stuff.'
US Team	*We had regular team meetings where we all sat in a conference room; one of the things that we did that I think worked very well is we reserved a team room and we had a dedicated team room where we could keep all of our materials and we met there weekly. In fact, in the beginning we met there twice a week.*
	Our team room, the 'Brisbane Room' was where we kept all the research material, the samples, where we held our meetings and where people went when they had a spare minute to work on the project.
SLT Comments	
	'As part of our support, when they asked for it, we gave each team their own room for brainstorming and working on the project and that seemed to be important to them to make them feel like a special group or team.'

6.3   Processes

This section explores the processes that were employed by the innovation teams. Beginning with the issue of the prioritisation of ideas and the ensuing conflict management (6.3.1); it also describes how the teams managed to harness the voice of the consumer (6.3.2) within their ideas. It further describes the teams' efforts to champion their ideas (6.3.3) in order to build internal support for them within the organisation.

6.3.1 Prioritising Ideas

Prioritising the ideas that were generated within both teams in the ISF project was problematic. It was the single activity that, according to the interviews, generated most disharmony and conflict within both teams.   But, it was especially stressful for the UK team, as their leader had made a very deliberate decision not to have any formal process by which to rank his team's ideas. He told his team that if they had sufficient passion for their idea to persevere with it; develop it and present it at the final meeting in US. He said that, purely on the basis of their commitment to it, he would support the idea being taken forward. For many on the UK team, this lack, not only of objective ranking, but of any type of ranking or screening process, was frustrating and failed to provide the type of structure and guidance they sought.   They felt that because no criteria had been set, that the group were potentially (and possibly inadvertently) treating large and small ideas with the same level of attention.   Moreover, because some of the ideas were very technical and specialist, many members of the team weren't appropriately qualified to judge their quality and relevance.

Separately, in a number of the interviews, at all levels in the organisation and in both research centres, it was acknowledged that the 'personalities in the UK were a lot stronger.'   This phenomenon materialised most in this context (deciding which ideas to promote and which to abandon) insofar as it was difficult to separate the ideas from their originator and it is posited in the interviews that some of the ideas made it through the process more on the strength of the personality and influence of their originator than purely on their own merit.

Theoretically, prioritisation of team ideas ought to have been a high priority for the leaders because they had a specific mandate to present back a maximum number of 9 new ideas. The UK team largely ignored this requirement and so they did not apply any upper limit to the number of ideas they presented. This lack of prioritisation certainly contributed to their ultimate loss of the competition (see Stan Lech's comments in Chapter Five). It is suggested by the judging team (SLT) that the strength of the personalities involved in the UK team made it difficult for Grist to formally rank his team's ideas and that this was a major flaw in his team's output. He failed to divorce the ideas from their originator and hence was unable to rank them, which made it very difficult for the organisation to deal with them afterwards and to assess which ones to prioritise.

Creating a rank order for their ideas was a problem too in the US team but for slightly different reasons. First, they had subscribed to a very rigid but objective and independent form of online research to evaluate the likely commercial potential of their ideas. This virtual research tool is very well established for multinational companies like GSK for providing some advance (e.g. financial planning) assumptions, like purchase intent, for product and line extensions that the company is planning. Many organisations of equivalent scale to GSK use the research approach and company involved. It compares the sales potential of new ideas to a database of benchmarks or norms that it has developed over many years of conducting similar research.

The strength of the research tool is in calculating potential incremental revenue opportunities for incremental product launches. It is not generally recognised as having the same level of accuracy when it is used on radical, breakthrough product concepts. Even its promoters acknowledge this shortcoming in the method as they concede that all new ideas are being compared to existing, often market-leading products. For this reason, some of the US team members who rated their own ideas more highly than the research did, were reluctant to accept its results. In querying the accuracy of the research, they claimed, possibly correctly, that the research instrument was not adequately sophisticated to measure the potential of their ideas. They felt that their ideas were too radical to be properly measured by conventional research techniques. This dispute led to a lot of conflict within the team.

Another source of conflict in this phase was a difference of opinion among a small number of US team members (and regularly voiced by one member in particular) about the company's genuine appetite for radical ideas. The initial brief had specifically told the group to seek out *Radical Ideas*. One of the principal researchers on the US group was convinced that although GSK had asked for radical ideas, that what they really wanted were near term, incremental, easy-to-implement ideas and hence this individual set themselves in strict opposition to any ideas that didn't fit that perception. This personal bias provided another barrier for the team in creating and progressing the radical agenda.

The table (Table 6.5) below contains the verbatim quotes from interview participants, specifically on this topic. No individual respondent is quoted more than once:

Table 6.5: Finding a Way to Prioritise the Ideas

Issue: Prioritisation of Ideas	Direct Quote
UK Team	*"If you fire up twelve or thirteen individuals who are supposed to be relatively creative and tell them to all go get their own ideas, then how do you then sit them down and argue which ideas do you take forward and which ideas do you leave behind? And do that in a way that divorces personalities and egos from it, is always quite tricky."*
	"This is where the system failed – because we had lots of very strong ideas. Some of them made it through because of the power of the personality rather than the strength of the idea itself." *"If somebody had genuine heart and enthusiasm for an idea, they were allowed to run with it, which actually I think is the right way of doing it because, remember, we're doing it from a scientific perspective. So some of the areas of science would not have been strengths of everyone on the team – so there had to be an element of trust."* *"There were ideas that were very personal to members of the group and largely went through purely on the strength of the passion the individual had for the idea. Personalities came into play too much."*

Issue: Prioritisation of Ideas	Direct Quote
US Team	"Although we were asked to try to come up with ideas or concepts which were totally radical, and spend a lot of time and effort in blue sky areas - I just didn't do it. I didn't think that the company would accept these idea and I did not support those concept and ideas items from others either." "Even after we had agreed to eliminate certain ideas from consideration – nevertheless, they kept coming back. People were unwilling to abandon their own ideas."
	"To have an effective team, you're bringing in people with creative abilities – and usually they're very passionate about what they do and not willing to let go of the ideas they care about. When you have passionate people promoting ideas – they are bound to come back again and again – and I think that's a healthy thing. I think we needed the passionate people." "Ideas kept coming back and this was a lack of formal leadership in ensuring the exit of the ideas and he wasn't adequately forceful. Scott should have played good-cop, bad-cop."
SLT	"In the UK, they had a few more rebels; a lot of rebels, a lot of free thinkers. They had more personality clashes, more problems – they were more out of control. If you look at the personalities – the personalities in the UK were a lot stronger." "If I'd been on the teams, I would have presented back a ranked order of ideas; I would have said 'my top three are....' But what we got instead, from the UK, was a long list of stuff with no prioritisation." "The edge went to the US team because it was structured and they had the data and they had the consumer piece. They had concrete consumer data and they ran the project in a metered, structured way."
Issue: Prioritisation of Ideas	Direct Quote
	"The UK team were unstructured people, who were unstructured with their time and tended to take on too much to begin with . Whereas in the US, the people were more structured and therefore less stretched. The people in the UK just tend to be a lot more free-spirited and less structured and that created a lot of conflict for them."
Close-Out Report	"If someone is passionate enough to want to present; then go for it, it is all about belief."

289

	"Struggled with commercial volumes estimates for ideas. Key is don't get hung up on them (numbers); focus on heartbeat and technical feasibility."

6.3.2 Harnessing the Voice of the Customer

Desouza et al. (2008) assert that organisations in today's competitive marketplace are increasingly innovating in partnership with their customers thereby subtly changing their innovation strategies from "innovating for customers" to "innovating with customers". In the ISF project, the issue of customer (or in this case, 'consumer') involvement in the innovation process was a particularly contentious one. This case study is laced with contradictions and differences both of opinion and interpretation on this issue.

The original project brief asks specifically for ideas that have been 'tested with consumers', implying that the teams will have had to go through at least one round of consumer research to be able to estimate the potential in-market value of their ideas. They are asked for a sales and market estimate, including some calculation of market share. So, there is an explicit requirement to conduct market research and to allow the voice of the consumer to have a say in which of the teams' ideas have the most market appeal.

In the US team, this mandate was taken very seriously and the team spent more than half of their total project time in the consumer research phase of the project. They had their ideas developed as concepts which were professionally written and illustrated. They then engaged in impressionistic, qualitative, focus-group research to finesse the elements of these concepts and identify the ideas with most consumer appeal. These focus groups also doubled as a learning experience for the team in which the members actually moderated the groups. This was then followed by early-stage, volumetric, online, quantitative research. Indeed, in naming them (the US team) the winner of the ISF initiative, the

President of R&D specifically alludes to the consumer research they conducted and cites it as the main reason that they had won.

Notwithstanding this view (which ultimately prevailed within the ISF context), there were other, differing opinions about this. The UK team leader felt strongly that he did not wish to dilute the focus on raw science by his group of qualified and experienced scientists. He did not wish to compromise any of the potentially ground-breaking and radical ideas they had by seeking consumer feedback within this phase of the process. He believed that the strength of his team was their scientific background and expertise and that they should really concentrate on this element to the exclusion of market research. His view was shared by the majority (although not all) of the SLT.

The question was asked, by a number of SLT members, about why would one bother to 'make a second rate market researcher out of a first-rate scientist?' Why should the scientists dabble in marketing when their primary role is to develop new and high quality ideas, rooted in science? It was also noted by the people who held this view that the ISF project had not been designed as a training exercise and hence it was a waste of the scientists' time to become engaged in market research to the extent that this was done in the US. Consequently, although the US team were officially the 'winners', a significant group of the SLT acknowledged that they believed the UK portfolio of ideas to have been more radical and certainly, more original. Moreover, more of the UK ideas are still in the GSK pipeline as of May 2011 than US ones.

It is worth noting that there was universal support, within both the teams and the SLT, for the idea of including consumers in the product development process but the debate really centred not around if but when (and to what extent) this should happen. The US team built the consumer feedback loop into the heart of their process while the UK deliberately made only a token effort at the very end. As already noted, the UK team's ideas were observed to be more radical than the US ones. The US ones were described as incremental, even predictable. While there may not be a causal relationship between the timing and level of consumer input and the degree of radicalness of the ideas developed; it is a noteworthy correlation. Because of their distinct focus on consumer research, it could be argued that the US ideas were 'market-led' while the UK ones represented more of a 'technology-push' approach. The US team spent 50% of their time, from March to July (2007), on the market research and concept development element of the project. The actual quotations below illustrate the sentiments of the stakeholders involved (Table 6.6).

Table 6.6: Harnessing the Voice of the Consumer

Issue and Source	Quote
Voice of the Customer UK Team	*"So I felt, as a bunch of scientists, that we should at least stick with our scientific heritage, try and look at science in a different way, come up with novel technical solutions rather than trying to pretend that we were commercial (marketing) folk when we weren't. So the thrust of what we were trying to do was very much get into the science, rather than doing a balance between science and consumer research."*
	"To me, it would have been much more beneficial if the commercial people would have worked on the ideas with us to flesh them out and help with prioritisation." *"Nigel (UK Team Leader) was very keen that we didn't stop ideating very early. And essentially given the timeline we had we could either choose to stop ideating about a month in and put everything through BuzzBack (online quantitative research provider used by US group) or we could choose to keep ideating and go with a slightly more qualitative than quantitative approach."*

Issue and Source	Quote
Voice of the Customer – US Team	*"It was really fun to watch some of the people on the team who had never done it (run focus groups). Because you could see the light bulb come on. It was truly an epiphany. As an observer of them I think I could see it in them almost more than some of them realised it in themselves."* *"Some of these people had never seen focus groups before. I don't think they really even had an idea how they were conducting prior to this so it was really good learning experience for a lot of people too. They got to do something that they just normally don't do in their job."*
	"Yes, before I had only some vague ideas about the consumer and how research is done but after the ISF I have more idea of what concept testing really is and how to test this and then how to read the test result ok. And then how to put the whole thing together and then get the consumer research knowing about how the consumer feel, how they respond to your concept. I'm a technical person and I'm able to come up with a modified concept immediately I hear the feedback and then, ask their input again and they respond again." *"Being out in the front room with consumers and talking with them, people found that to be extremely valuable, so much more beneficial to their understanding of the idea of when they are talking with them human being to human being, person to person and getting the scoop right from the horses' mouth."*
Voice of the Customer - SLT	*"I didn't find this refreshing because a lot of money and a lot of effort went into this marketing effort. I didn't think this was appropriate because this was not a training exercise. As scientists, they ought to have focussed exclusively on the science and not got bogged down in marketing."* *"The edge went to the US team because it was structured and they had the data and they had the consumer piece. They had concrete consumer data and they ran the project in a metered, structured way."* *"What I saw was that a lot of effort had been directed towards consumers – what I was loking for was far more of the scientists doing science! And far less of scientists trying to run focus groups."* *"A lot of the US stuff had been done before and a lot of the UK stuff was completely off the wall, in a good way, new – just what we wanted."*
Voice of the Customer – Close out Report	*"A chance to communicate with consumers, challenging but great experience."*

6.4   Networks: Internal and External

The working style and task approach of the two teams was characterised by considerable variation between the US and the UK, but nowhere was this divergence more evident than in their use or non-use of networks. Informal networks within the firm, as well as the network of external linkages that they created and managed for this project, had an important bearing on the strength, the diversity as well as the kind of innovative activity conducted by the UK group in particular. The UK team actively engaged in discussions with internal experts and external collaboration partners while the US team-members, almost exclusively, confined themselves to desk research. Moreover, for the US team, their meetings were closed, they did not invite people in to share their ideas nor did they engage in any purposeful outreach programme in developing their ideas.

One of the very first actions of the UK team leader was to brief the project out to an innovation intermediary. He contacted Bufton Consulting, an innovation agency, specialising in open innovation, working in the healthcare field in the UK but with links internationally, and gave them the same brief he had been given as the ISF project mandate. This immediately gave the UK team access to novel ideas and technologies which may not have occurred to the team members were it not for the Bufton connection. He also encouraged his team to go out and meet with experts in the fields in which they were interested and this did happen.   The Sleep Laboratory in the Surrey Sleep Research Centre became a contributor to some of the ideas proposed by the UK team.

The interviews show that the UK team adopted an open approach where they actively sought fresh input and novel ideas from external experts and partners. They forged alliances, they met with experts and they invited intermediaries in to help them identify potentially interesting new technologies and therapy areas. The US team, on the other hand, were more internally focussed. They had closed meetings in which they exchanged their own ideas and they logged their ideas using a computer programme called ThinkTank. This programme allowed them to post the ideas, to classify them under certain headings, to rank them and to vote on them. One of the US team also suggests that the US team found it easy to generate a high number of ideas because many people went into these idea generation meetings with a number of old ideas that had already been discussed prior to this initiative. While the US team focused their attention on getting consumer input into their ideas, the UK team concentrated on getting external, expert input into the generation of theirs (Table 6.7).

Table 6.7: Building Networks

Issue and Source	Quote
Forging Networks– UK Team	*'So the philosophy was that rather than have a process whereby you go through different stats (market statistics) to try and pull ideas together, is you create an environment where people have the opportunity to read about new areas, talk to experts in different areas, interact with different people such that they can generate threads for their own development.'* *'We were inviting in the external experts; we were highlighting, to each other, interest in sources of new scientific development and trying to create an environment where our people would explore new areas and find out something they thought was of interest and then develop threads'*
	'So I felt, as a bunch of scientists, that we should at least stick with our scientific heritage, try and look at science in a different way, come up with novel technical solutions. We wanted to get into the science, talk to experts, see what's exciting out there and bring it back and discuss it in our team.'
Using networks – US Team	*First step was coming up with ideas and we used a software package called group systems – which facilitates the building and adding of ideas.* *Initially what we did was we all just sat around the table and tossed ideas out and, you know, kind of wild and wacky stuff, the crazier the better.*
	We did not consult anyone outside the company in developing the ideas; we did it all within the group with one exception when we needed expert opinion on ozone
	Generating the idea was pretty good actually. Basically many of us came in with old ideas so that's why we would generate over 300 ideas so easily.

SLT Comments	
	The behavioural piece, we were looking for people that were very open to radical ideas, we were looking for the people that had previously suggested to us ideas that didn't fit in the current business that we're in, we were looking for people on the R&D side that had least a string of patents or applications, so we know that there's not only the initial thought, but also the follow-through. *We would have far preferred to see the US team engage with experts instead of engaging with consumers – they might have got more original ideas that way.*

6.4.1 Championing the Ideas

One of the key features of the design of the ISF project was that it purposefully dispensed with the need for the team, or specifically the team leader, to make any regular or even interim progress report to senior management. This applied for the full nine-month duration of the project. It was seen as an emancipatory measure, to free up the team, from 'red tape', to concentrate on the science and to pursue appropriate, novel ideas without the conventional reporting constraints that would normally apply to innovation teams working in GSK. It was considered by the SLT that the paperwork involved in the conventional reporting protocols were too onerous for some of the more creative R&D people and by removing them; these people may have a unique opportunity to flourish.

The teams were required to report in only two circumstances: a) if they needed specific resources and b) if they over spent their budget. Both teams approached senior management for the allocation of a dedicated creative space; and both secured the use of a special room at their respective sites. They were able to use this room to store samples, competitor products and prototypes and it was also used as the dedicated space for people to work on the project either on their own or in sub-groups. The rooms also became the places where each team would hold their meetings. Neither team needed to request additional funding.

In complying with this element of the project structure, the two teams, once again, adopted vastly differing strategies. The US team adopted an approach where they didn't formally report progress, but informally the team leader stayed in regular contact with the SLT. He briefed all the VP's (on the SLT) during the project to sensitise them to the types of ideas that were surfacing in his team; he also wanted to make sure that they were looking in the right areas and territories. He was also anxious that his team didn't put a lot of effort into areas that might have been explored before. He correctly identified that the organisation did not, at that time, have a knowledge bank or knowledge management system where ideas, which are currently under review or ones that have previously been investigated, are filed. Hence, it was entirely possible that his team might spend a lot of time and resource on something that had already been considered and discarded. By doing this extensive but informal consultation process, he was conditioning the judging panel so that they knew what to expect from the US team and there 'would be no surprises'.

Anticipating possible future objections to the team's output, the US team leader also spent a lot of time speaking with the (US in-country) marketing teams to make sure that he was following the appropriate guidelines for how concepts should be written and ideas presented for research within GSK. As was previously noted, there were no commercial experts on either team and the US team leader wanted to make sure that he was following the appropriate guidelines for bringing the commercial ideas to life. Hence, he consulted extensively with the US commercial teams and sought and used their advice. He ascertained who the preferred suppliers were for these services (illustration and copywriting) and he used the same vendors to help develop and render the concepts from his team. In this way, he hoped to be able to overcome any possible objection from the judging panel about the quality or the validity of his team's work. He did the same for the research providers for both the qualitative and quantitative phases and hence the US team employed not just the same research techniques to validate their ideas but they used the same companies that the commercial arm of GSK would have used themselves.

In contrast to the US team leader's approach to internal boundary spanning (or bridging-ties); the UK team leader did not engage in any discussions with the SLT or any other internal group. His approach was to keep everything under wraps so that the SLT would be 'pleasantly surprised' when they saw the ideas for the first time at the presentation in Manhattan. Below (Table 6.8) are some of the direct quotes from the team members and the other stakeholders as they relate to this issue.

Table 6.8: Innovation Champions in ISF

Issue and Source	Quote
Championing Innovation – UK Team	*"Nigel (UK team leader) said that 'this is an R&D initiative' – he encouraged scientific type ideas but not necessarily commercial ones. He had a scientific approach rather than one which was consumer driven. He didn't talk to anyone in senior management about the ideas that were on the table. We were sailing blind."* *"Some of our ideas were on the brand, Aquafresh and we didn't want to show them to (the team on) Aquafresh too early in case they didn't like them, and then when we present them, we present an issue."* *"We wanted to make an impact and so we didn't want. Them (SLT) to know in advance what was coming. I guess we had faith in the ideas."*
	'I think they deliberately stayed away and the whole point of it was 'what would these guys do in a vacuum with a little money and just a brief?' So leaving us alone was a vote of confidence in us and our ability to deliver ideas.'

Issue and Source	Quote
Championing Innovation – US Team	"He (US team leader) did lots of sensitising the management to what ideas and areas we were thinking of – but he didn't get us involved with these bi-laterals - Scott (team leader) did it all and got the right tools, using the right people and getting quantitative data as well."

"We had quite a few of them (ideas for GSK brands) and we suspected that maybe these were some ideas that had already been looked at. So rather than spin our wheels and go back reinventing the wheel, we did convene a meeting with the commercial and R&D Futures Heads (the SLT and their commercial counterparts), and said here are some ideas that are being batted around, we think there is some value to them for these reasons, would they be on strategy? Would they be off base from what you consider? ... Just to make sure we weren't reinventing the wheel and going down a path that they already had data to say this is a non-starter for us."

"We knew he (US team leader) was in dialogue with senior management and this ensured that they knew and approved of what we were doing – this was very reassuring for us." |
| Championing Innovation - SLT | "My observation was it was all very underground when it was going on. It appeared to be quite underground and I had thought even though SLT's brief was, you were not to step in and manage this, let them be. I still thought, I had expected that people would give me a shout and just tell me where they're at and no one did. And I have to say it was an element of surprise for me."

"When we first set it up, we told them right away we did not want to be 'in their underwear', you know. We specifically told them why we picked them, they said you are the best, the brightest and creative people we have and we gave them a lot of confidence saying – we know you can deliver this."

"Along the way I think I checked in with Scott twice and I think Nigel just once. But Scott would often show me some ideas and I would go – that's interesting. I would give a little bit of guidance, but not much, but they knew that we were interested but the interest was not meddling." |

6.5   Team Leadership

The themes discussed so far; team structure; team processes and building networks have all been drawn from the lived experience of the participants in the project. In each case, while there were some necessary components built into the programme, the overall ISF project design was found to be imperfect in some significant way on all counts. However, all of these individual issues are connected to an overall theme of team leadership. One key element that makes this particular case study such an intriguing and revelatory one is that the two assigned leaders took such divergent positions on how they would perform their leadership role.

From the transcripts of the team members' experiences of the project, it is clear that the approach of the leader is paramount. The influence of the leader is difficult to overstate in this case study. That the two individuals interpreted this role in such vastly different ways, adds to the texture and richness of this study. Although one team was declared a winner; this was a relatively hollow accolade when most of the judging panel actually sided with the 'losing' team. Moreover, subsequent career moves within the organisation have seen a significant promotion for the leader of the 'losing' team while the leader of the 'winning' team has not seen any equivalent promotion.

The case study is described in detail in Chapter Five, however, it is worth recapping on a couple of the significant elements with regard to leadership and how it was manifest within the teams; because it was the differences in the leadership style and approach that lend this study a particular fascination.

Within the ISF initiative, the selection of the team leaders was done with deliberate care. However, the choice of the UK team leader attracted a lot of controversy and outright opposition with even the SVP of R&D acknowledging that this particular selection 'was a bit of an experiment'.

The UK team leader chose a hands-off approach. He never brought the full team (physically) together; he didn't channel the work; he didn't assign people into sub-teams and match the tasks to their respective strengths; he didn't engage in any boundary-spanning or coalition building for the ideas; he didn't adhere to the terms of the brief; he didn't objectively prioritise the ideas but just allowed people run with the ones that appealed to them most; he didn't manage the research piece effectively and when people on his team needed direction or guidance, he failed to provide it. He didn't channel the work or provide feedback on the quality of the ideas. One member of his team experienced a 'significant, stress-related, health incident', which kept them out of work for over 6 months afterwards and was widely attributed to the stress that arose from being on the UK team. But despite these apparent failings of conventional team-leadership, the UK team-leader was absolutely true to the twin notions that creativity requires freedom to explore ideas and that innovation needs to have an element of chaos to emerge. He also exhibited a style of Pygmalion type leadership by continually expecting his team to deliver the high levels of creativity that he originally thought they possessed or, at least, hoped they did.

Moreover, the ideas his team produced were widely considered to be more imaginative, creative and disruptive than those of their US rivals. A key pillar of his approach was that he refused to interrupt the scientific focus of the ideas to engage in any consumer validation or market research. He believed that the project offered 'a bunch of scientists' an opportunity to explore scientific ideas and he remained faithful to this till the end. He accepted that chaos is an essential part of the fuzzy-front-end of innovation and he allowed, or even encouraged, chaos to override structure.

He, effectively, defied the original brief for his team and ignored the requirement to select a small (eight) number of well researched ideas in favour of devoting more time to the science (or applying idea generation to science) and ultimately coming up with a high number of untested ideas. He defied the mandate in two ways. First, he didn't carry out any meaningful market research on any of the ideas and; second (but related), he did not rank them and prioritise the top scoring ideas. Instead, he merely presented a large palette of ideas with no indication of which might be the ones with highest potential.

A vastly different approach was favoured by his counterpart; the US team-leader. A former career project-manager, his well-organised style underscored his commitment to timelines and processes. He managed to get his team to meet a couple of times a week, he introduced dedicated innovation software to the process so that his team could work on, register and even vote on ideas in real time regardless of their location. He planned out the phases of the project and made sure his team were kept on track. He channelled their work effectively and he was totally faithful to the original brief. Not only did he conduct an exemplary research element to the programme, he also integrated some useful learning opportunities for the R&D team to interact, often for the first time, with consumers (although, this last piece proved highly controversial). He delivered on all the objectives and he used his corporate experience, by arranging informal meetings with all the key stakeholders, to ensure that the SLT were prepared for and positively disposed to the ideas that his team had developed. His highly organised approach ensured that his team won the competition. But notwithstanding being declared the winners, some questions remained about the quality, ambition and even originality of the ideas his team came up with. He was also heavily criticised for spending so much of the project time in the research process; there was a suggestion that bringing the consumers into the process so early and so intensively blunted the creativity of the ideas. He was also criticised for making the process too technology-driven and having too much bureaucratic process. This latter comment is a phenomenon already recognised in the literature, specifically noted by Loewe and Chen (2007); 'But sometimes, an overly eager project manager intent on making sure that everything gets done on schedule will let the calendar take precedence over the content and the quality of the outputs.'

Below are some direct quotes from both team leaders and members, provided alongside quotes from the judging team and some lines from the official project close out report (Table 6.9).

Table 6.9: Team Leadership in ISF

Issue and Source	Direct Quote
Team Leadership – UK Team	"The UK approach was chaotic with process largely absent. It was like a playground where you could do exactly what you wanted."
	"We were all smart enough to realise that unless you have some structure in place there was a risk that the chaos would have very little output. So most of us wanted (stage) gates to aim at and deadlines to hit. And, we never got them."
	"Nigel used very little framework for running the team. He never set the scene but simply allowed everything to be completely flexible. While innovation needs chaos – maybe the people involved don't. And, a bunch of scientists generally want some structure. I think it was a little bit chaotic and over the top."
	"I mean, the US had a much more structured approach, they had dates by which they had to stop having ideas for example, whereas we were effectively having ideas right up to a couple of days before (the presentation) and some stuff made it into our final presentation that had only come to the tables – well in fact there was some stuff that went into the presentation that none of us had actually seen other than the person who presented it. So our process was wide open."
Team Leadership – US Team	"When I kicked off the meeting, I had a very clear vision in my mind of what we wanted to be in a position to present to SLT seven months later. And it wasn't just some ideas that have been bounced off a couple of consumers perhaps, or bounced off internal people. I wanted to be able to bring forward quantitatively tested concepts of new product ideas that we had actually thought through on the technical side and had clear approaches on how we would go about the whole thing technically. But I wanted to make sure we had the consumer heartbeat established to the point that we had some quantitative concept consumer test results on that."
Issue and Source	Direct Quote
	"Scott's planning was essential, he had an end-game, he had a timeline. Creativity is the well-spring of innovation but you have to move it forward and Scott's processes really helped move it along comfortably."
	"Scott created a high comfort level – people were very open, transparent, nobody was uncomfortable. He was very neutral – he was very good in terms of the psychology – people felt comfortable."
	"We wanted to make sure that the organisation's curiosity was

	met and we did what we called an ISF fare and we set up all of our posters and concepts and learning out on the patio at the GSK café here and had an open fare (exhibition) where we actually displayed everything that we did and people were able to walk through, we had each of the sets of concepts manned by one of our ISF members and people cold stop by and look at the concept and read it, ask questions about what it was about."
Team Leadership – SLT	"I had concerns that they hadn't got team-working skills; we had no leader, no coordinator and I raised strong objections to the point that my objection was noted. I was afraid we were setting them up for failure – particularly the UK team. My objection was noted but disregarded." "One of our people, who made the main presentation for UK team, experienced a pretty serious medical incident afterwards that kept him out of the business for six months – and I wonder how much did the ISF project contribute to his condition? If there had been a better structure in the UK, I wonder would he have experienced this condition?"

Issue and Source	Direct Quote
	"If we had commercial members as part of the team, we might have been able to prioritise the ideas afterwards a little better – because now we have no modus operandi for moving the ideas forward. We've had to sift and sort without really having all the data necessary to make the right decisions."
Team Leadership – **Close Out Report**	• *Leaders owned logistics* • *No right or wrong way of doing things* • *UK team ignored guidelines* • *US team focused on timeline and delivery*

6.6   Summary

This chapter has taken the interview transcripts and highlighted the themes that emerged from the lived experience of the team members. It focuses on the team-level and individual-level experiences to identify issues which may possibly be converted into framework guidelines to help increase the efficacy of future innovation management programmes in large, global organisations. The research objective was to analyse, in detail, how an innovation project actually unfolded within GSK and to isolate the key factors which are conducive to innovation across the three phases of the innovation value chain (Hansen and Birkinshaw, 2007).

This case study reflects the complexity known to characterise the innovation process. It seems apparent, from this case at least, that there are four specific dimensions that ought to be addressed in the innovation process for companies wishing, not only, to create an infrastructure conducive to innovation but also wishing to get tangible innovation outcomes from such initiatives. These four dimensions are structure, process, networks and leadership. The findings from the case study for each of the themes will now be discussed.

6.6.1 Team and Project Structure

A first important finding of the case is that basic, hygiene-factors ought to be anticipated and built in to the planning phase of initiatives like this one. Such factors are reported, in this case, to have an adverse effect on the experiences of the individual team members. The case data also suggests that had this issue been managed differently, the outcomes of the projects might have been superior; although, this cannot be stated with any real authority. These fundamental issues relate to how these teams and programmes are set-up and structured. The key issues around structure that emerged as significant in terms of the impact they had on the team were the appropriate allocation of time to do the work required. Also, issues like "squaring-off" the project with the team-members' line managers were thought to be crucially important. The size of team is also a factor as is the duration of the project and whether the entire team, or even just the leader, are devoted to the project full time. Having a dedicated, appropriate space to innovate was also reported to have a bearing on the experiences of the teams.

What this study has shown, in summary, is that the majority of the interviewees, following their experience with the project, believed that it would have been better to have had smaller, tighter and (full-time) dedicated teams. They suggest that teams of between five to eight members would have been preferable to teams of twelve or thirteen members.

The interviewees also make the point that the organisation failed to communicate either the purpose or the profile of the project to the line managers of the team members. This had the regrettable consequence that some of the team members were made to feel that working on the ISF project was an almost subversive activity as they had not received any encouragement or mandate to work on it from their direct line manager.

However, despite being raised in many interviews, it remains debatable just how important these issues of structure really were. In terms of outcomes, both teams satisfied the brief and delivered what was required, albeit with considerable variation. It seems that, at best, it can be asserted that many of the team members report that they would have been able to perform to a higher standard had the company managed to anticipate and manage these issues better. These cautions should serve as a reminder to managers charged with setting up innovation programmes that these structural elements are significant and they warrant spending some time to get right.

6.6.2 Issues of Process

The second tier of issues that need to be addressed in setting up innovation teams lend themselves to the description of process issues. This refers to issues around how the team manages the phases of the innovation journey. Under this broad heading come specific issues like how the team rank and screen their ideas; how they develop them into testable, research-ready concepts and how they harness the voice of the ultimate customer into the innovation process.

In this case, neither team had a robust or definitive way of screening their ideas and this caused team conflict in both teams. Although the UK team applied the personal passion of the individual team member as the first filter of the quality of their ideas; this caused some disquiet amongst the rest of the team who worried that without any additional criteria, an inordinate amount of time might be devoted to ideas which may be of very low commercial potential.

In the US, the team developed a screening process with two phases; an internal ranking and an external one. First, they used the software tool, ThinkTank to vote on their entire portfolio of ideas, taking only the top ranked ones forward to qualitative research. After that, they used a volumetric tool called BuzzBack to assess, using an online consumer panel, which of the rest of the ideas had most appeal with the target market at which they were aimed. This way, they had a good sense of the potential in-market value of each of their ideas. But nevertheless, conflict emerged within the team, for two reasons. The first reason was that some among the team felt that their own, individual ideas were truly radical and therefore that the testing mechanisms (such as BuzzBack) used lacked the subtlety to be able to appreciate the full potential of their ideas.

The second issue which exacerbated the team conflict in US was the very passionate view held by one team member, in particular, that despite what the company said in its briefing about the quest for radical ideas; she held that GSK were not really serious about this and had an appetite only for the most basic upgrades to their product portfolio. Hence, this team member argued constantly against the inclusion of any ideas that could be termed radical. Team conflict was unnecessarily high in both teams although it arose for different reasons within each team.

The issue of the voice of the customer was handled very differently by both teams and it will be argued later that this had a significant influence on the type and dimension of the innovations proposed by each team. In the UK, the team leader felt that the opportunity to get experienced scientists to immerse themselves in the scientific data surrounding new ideas was more beneficial than using the project time to canvass the views of potential future customers. He argued strongly in favour of allowing the scientific experts engage with the science, create and build new ideas and, when the time was right, he felt that marketing people could evaluate the market potential of those ideas. In this sense, he ensured that his team actively spent all the ISF project time in the idea generation phase of the initiative. They never really moved into any further elements of research or testing. The UK team leader showed a total commitment to the science element of the idea generation phase and he eschewed the notion of asking consumers for their reaction to the ideas (thinking they would only dilute them), wanting instead to maximise the time he could keep his scientists refining the science of their ideas.

In contrast, the US team leader spent only a quarter of the project time, generating the ideas. The US team started to research their ideas first through qualitative research using focus groups and ultimately through quantitative online surveys. The US team were active in all the stages of the innovation process. While the UK placed their faith in the ability of their scientists to unlock some great ideas by focusing on the technology, the US had equal faith in the role of the consumers in developing high potential new ideas. For three quarters of the duration of the ISF project, the US team were engaged in the process of consumer research in its various phases. The US team's process of validating their ideas through consumer research paid dividends for them, in terms of the competition outcome. It meant that they were ultimately presenting a limited palette of pre-validated ideas, all of which were researched and which had in-market values attached to them. They were presented using the appropriate concept writing format; the right type of research numbers with which the senior management team were used to looking at commercial opportunities and these features made them easier to assimilate into the prevailing business model than the UK ideas which were still relatively raw by comparison. The leader of the SLT who designated them as the winners specifically alluded to the fact that they had done the consumer research as the critical element which swung the decision in their favour.

Nevertheless, the rest of the SLT members were not convinced that the US were worthy winners. In their interviews, they described the US ideas as predictable and incremental. Conversely, they saw the ideas presented by the

UK team as being more original and radical. In the end, the competition or tournament was not conclusive. It had an official winner; the US and an unofficial winner; the UK.

6.6.3 Network Issues

A third element that features as instrumental in the process is the extent to which the team and the leader used networks both within and outside the organisation and how leaders champion their ideas internally with senior stakeholders to increase the likelihood of their adoption after the project has concluded.

Innovation champions play a vital role in protecting ideas from premature evaluation and in persuading senior managers of their merit. The US team leader maintained a constant but informal dialogue with the SLT members so that they were aware of the nature and type of the ideas being developed on that team. Moreover, the US team leader made sure that his team were not wasting team resources on ideas which had already been evaluated in some prior R&D programme. The US team leader sensitised the internal audience for the ideas his team were working upon for the competition.

By contrast, with considerably less project management experience under his belt, the UK team leader, chose not to communicate his team's ideas to the SLT, he wanted the ideas to be a 'surprise' on the day of the final competition presentation. This strategy did not work out so well for him and attracted some negative comments from members of the SLT who expected to be consulted during the process, notwithstanding the provision in the project set up which freed the team leaders from any reporting constraints.

Internal experts can often add understanding and insight to ideas which can strengthen them immeasurably. The UK team identified, pursued and consulted with internal experts in certain therapy areas. The US team did not connect with anyone from within GSK on their ideas. GSK has over 100,000 people on the staff and it tries to promote internal connectivity between them in order to share and develop knowledge within the company. By neglecting to reach out to possible experts within GSK, the US team were missing an opportunity to develop their ideas.

But for organisations seeking radical innovation, external networks seem more significant, than internal ones, in achieving this objective. The UK team excelled at this part of the project. They approached Bufton Consulting, an open innovation intermediary, and effectively sub-contracted the entire project by providing Bufton with the brief they were working to. Bufton consulted with their client base of inventors, research institutes, academics, start-up technology companies and reverted to the UK team with some valuable external contacts. The UK team also engaged with the Continua Alliance (an Open Innovation Network specialising in connected healthcare) and this connection provided them with a number of potential leads from which many of their ideas eventually emerged. Teams which purposefully manage their external networks appear, in this study, more likely to find novel and useful ideas to bring into their innovation programme. Many of the UK ideas relied on external technologies and originally sprang from the external, open contacts the UK made from early on in the project. This external focus was largely driven by the team leader who, made useful eternal introductions for his team.

The US team, by contrast, neglected this networking element of the idea generation phase. They failed to engage with anyone outside their own team and this omission was obviously a limiting factor in the range of ideas that they ultimately recommended. In one instance, they did consult with an external expert (in ozone technology) but this was really to fine-tune an idea upon which they were already working. The most common criticism of the US' portfolio of ideas was that they were predictable and incremental and they lacked the vision of those suggested by the UK team. It seems that one reason for the more limited range of ideas was that the US team did not engage in networks wither within GSK or outside it.

6.6.4 Leadership Issues

Finally, and this is possibly the strongest finding of the fieldwork, leadership is a critical component to the direction, the momentum and the outcomes of innovation programmes. In the ISF story, both leaders achieved a version of success but they did so in very different ways. One leader led the innovation charge by systematising the procedure and minimising the opportunity for chaos to develop. The other preferred a looser, more flexible, highly open and networked approach. This study points to an issue that remains unresolved in the literature; can the same type of leadership be equally effective along the three phases of the innovation value chain? Similarly, can the type of leadership that yields incremental innovation also be effective in the quest for radical innovation?

As noted, there is rich contrast in the leadership approaches adopted by both team leaders. The UK team leader stuck rigidly to his philosophy that 'the only process is there is no process'. He was also absolutely faithful to the notion of the scientists sticking to the science and not dabbling in market research. Therefore, he defied the brief by not evaluating the commercial potential of the team's ideas through consumer research. He did this purposefully in order to spend the additional time being immersed in the science and technology. The UK team leader also brokered valuable introductions for his team and forged connections both inside and outside the organisation. This networking approach had the effect of shepherding in more creative, original and novel ideas into the UK portfolio of ideas.

The next chapter takes up some of these issues further. This chapter has looked at the experiences of the ISF team and has used them to explore the themes that naturally emerged from their involvement in the project. Consequently, the starting point for this chapter is the team-members' experience. The starting point for the next chapter is the innovation process and over that is laid the teams' experiences.

Chapter Seven

An Analysis of the Project Outcomes; Evaluating the Ideas

7.1 Introduction

This chapter has three objectives: first, to discuss the outcomes (in terms of the NPD ideas proposed) from the project teams and draw conclusions about the type, number and quality of the ideas or concepts developed by each team. Second, to classify the concepts on a spectrum of radical to incremental and, third, to discuss the project outcomes and processes through the lens of the Innovation Value Chain and to attempt to connect the variation in approaches taken by each team to its three phases.

This chapter explores the ISF case systemically, or holistically and analyses the variation in project processes and resulting outcomes that was described in Chapter Five. This chapter will look at the objectives of the ISF initiative and follow this through right up to the outputs. The analysis provided here looks beyond the sentiments, experiences and opinions expressed in the interviews and provide an analysis of the programme outcomes. Moreover, the level of analysis in this chapter now shifts beyond the teams and focuses on the innovation project itself. It attempts to match the case data on this project with the theory surrounding the innovation process and, using this backdrop, to explain the wide variation in outcomes for the two teams. The chapter will begin with an examination of the differences between the two teams' outcomes and then discuss differences in their processes and attempt to identify the strength of the link between the two.

7.2   Description of the Outcomes from each Team

In order to understand the project outcomes using existing literature, it is necessary to revert back to the case study data itself and present exactly what those tangible outcomes were. Both teams were given identical instructions; to come up with nine novel, 'radical new product ideas' for the business. Six of the ideas were, ideally required to be aligned to the global brands currently marketed by GSK. Hence from these ideas, the SLT were hoping to get novel, high-potential ideas in therapy areas or categories with which the organisation was already familiar and successful. To respond to this element of the SLT brief would have required the teams to look at new technologies, new 'active ingredients', new delivery systems (oral, nasal, suppository, transdermal etc) , new formats (i.e. liquids, gel-caps, tablets, transdermal formats, sprays etc) or even new routes to market, new business models, new service ideas or new distribution models within the wide framework of the therapy areas and health and wellness markets in whch the company already operated.

The brief also asked for three additional ideas outside the current 'footprint' of the company i.e. ideas in new areas, new therapeutic categories in which the business was not currently operating. The specific chart from the briefing pack is shown below and makes these requirements very clear:

Figure 7.1: Slide from the ISF Briefing Meeting

> **Output**
>
> ° Aspirationally, one big idea for each of our current Future/Weight Control groups (Big Incremental Opportunity)
>
> ° Aspirationally, three big ideas which fall outside of our current Future/WC model (Big Incremental Opportunities)
>
> ° Launch 2009-2011
>
> ° The quality of the idea is the over-riding consideration; not the number of ideas
>
> ° Highly professional and sparkling presentation

Source: ISF Briefing Pack – Internal Documents.

This study has already provided a situational context from the perspective of the organisation in which the initiative took place, GSK. But what has not yet been discussed is the situational context for this programme within the innovation process itself. Had the teams been totally faithful to their original brief, their final presentation should have comprised a set of novel, commercial ideas in the therapeutic categories (Figure 7.1). The table below reflects their mandate; one new idea for each of the organisation's six core categories and then three additional concepts from outside the footprint of the existing business:

Table 7.1: The SLT Mandate to Each of the ISF Teams

Category	Number of Ideas Requested by SLT
Smoking Cessation	1
Aquafresh (oralcare brand)	1
Sensodyne (brand specialising in relief of dentine sensitivity)	1

Weight Control (alli – the main GSK band in this category)	1
Denture Care (Polident and Poligrip being the big GSK brands in this market.)	1
Pain Management (Panadol being the major GSK brand)	1
Other – ouside those therary areas	3
Total number of new ideas required from each team	9

In attempting to explain the wide variation in the outputs of both of these teams, it is worthwhile paying close attention to exactly what each team produced. At the final ISF presentation, which was the ultimate showcase for their outputs, the US team presented seven concepts to the SLT (as opposed to the nine that had been requested). They really arrived at that number based on two considerations. Firstly, these seven were their top performing concepts in research and, secondly, they knew that in the presentation itself, they had between three and four hours to present their portfolio of ideas. In order to do justice to each idea, especially to be able to show video footage of the consumer reaction to the ideas and to present their research results, they allocated thirty minutes in the final presentation for each idea and settled on seven as the ideal number to try to communicate within their time slot.

The UK team, for a number of reasons, did not prioritise their ideas. Moreover, they had no research results to present except of the most preliminary kind. Hence, they allowed their people present as many ideas as they could and they allocated their time by person rather than by concept. They presented 27 ideas in total; some more finished than others. Neither team really conformed strictly to the brief; neither side, for instance, found any new ideas for the denture care category. Table 7.1 below lays out the teams ideas against the quotas requested in the brief.

Table 7.2: Ideas Delivered Compared to Original Brief

	Actual Distribution of UK Ideas Presented	Actual Distribution of US Ideas Presented

	SLT Mandate		
Smoking Cessation	1	1	1
Aquafresh	1	1	
Sensodyne	1		1
Weight Control	1	1	
Denture Care	1		
Pain Management	1		1
Other – ouside those therary areas	3	24	4
Total number of new ideas from each team	Required 9	Delivered 27	Delivered 7

In analysing the ideas further, another layer of variation of outcomes between the teams becomes evident and this relates to the classification or typology of the individual ideas. The brief asked for 'radical' ideas. Indeed, the GSK R&D team overall had restyled their department logo to include the terminology 'Radical R&D' within their nomenclature and internal company branding. In order to discuss in more depth the relative positioning of the ISF ideas on the radical: incremental spectrum, it is worth pointing out how GSK classifies radical innovation within its internal project management processes.

In assessing its portfolio of projects, GSK looks at two principal elements within specific NPD projects; the technology required to deliver the idea and the market (customer segment) for which it is intended. If both the technology required and the target market are outside the current operating footprint of the organisation, then such an innovation is classified as 'radical' or 'breakthrough'. If both the technology and the target market or customer segment are within the current scope of operations; such an innovation would be internally classified as 'incremental'. The following paragraphs provide some examples of NPD ideas that fall either into or between these two opposing categories.

7.3 Innovation in ISF – Incremental or Radical

In reviewing the individual ideas from both teams and attempting to assess them in terms of either radical or incremental ideas, adapting the broad criteria indicated above, they have been classified into four categories which reflect the areas of both technological and marketing uncertainty.

a) 'Incremental' innovation captures those ideas which use the same broad technology, mode of action, science or active ingredient within an existing brand. This suggests that the idea is a minor line extension within a known market and using known and familiar technology. A 'mist pen' for Nicorette (see below) is an example of this category where the market targetted is familiar to the organisation; smoking cessation, and the active ingredient also familiar: nicotine. But the delivery mechanism, a spray mist device, is new. The main nicotine replacement therapy (NRT) brands are already available in gums, lozenges and even nasal sprays and this idea is for a new NRT format; a spray pen (see below).

Figure 7.2: The Commit Mist Pen

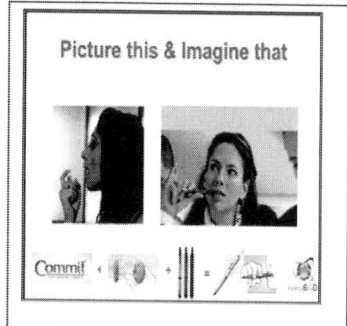

Source: US Final Presentation ISF Project.

b) The second classification is 'New brand but known science'. This is the category that attracted the fewest ideas. It proposes that GSK brings existing science into a new therapy area that would warrant the creation of a new brand. An example of this was an idea of using Blackcurrant technology (from Ribena) to develop skin-care products based on possible skin care benefits of the rich anto-oxidant concentration in Blackcurrants.

Figure 7.3: Ribena Skincare Idea

Beauty from within Opportunity
– Ambient Beverage Shot

- Great tasting all-natural Blackcurrant beverage smoothie shot. clinically proven to preserve Skin Moisture by upto 20%

- Containing the natural active ingredients:
  - EFA's extracted from Blackcurrant seeds (GLA)
  - Antioxidants from blackcurrant skin, and polyphenols from green tea
  - Vitamin E

- How does it work?
  - Drink 2 x 100g bottles per day
  - Can be carried conveniently ambient stable & small
  - Measurable increase in skin moisture after 6 months

So Why the Blackcurrants?

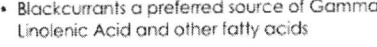

- Ribena – a blackcurrant drink

  - >400 tonnes p/a blackcurrant seed (UK only)
  - >1200 tonnes p/a blackcurrant pomace (UK only)

- Blackcurrants a preferred source of Gamma Linolenic Acid and other fatty acids

- Blackcurrant Cultivar Selection

- Traceability

Source: UK Final Presentation, ISF Project.

c) 'New science to known market' describes the process of adding new and better technology to existing brands. Self-warming toothpaste for Aquafresh or Sensodyne (see below) is an example of this category. A warm toothpaste would have more efficacy in killing bacteria. Toothpaste is a familiar market for GSK where they are the number-two, global competitor; but, self-warming toothpaste represents a new technology for that market.

Figure 7.4: Self-Warming Toothpaste

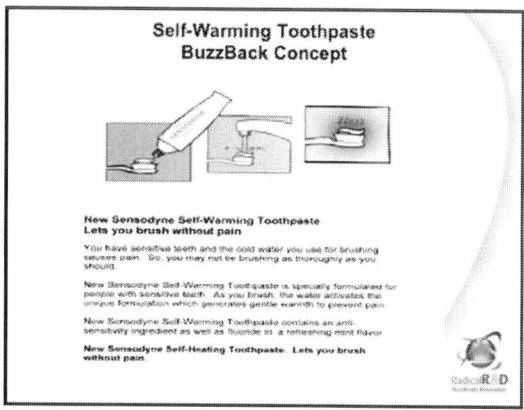

Source: US Final Presentation, ISF Project.

d) 'Radical' is the description given to ideas which take the company both into new markets and into unfamiliar technology. These ideas propose the development of new brands (new marketing competence) and new scientific areas. An example of this idea is the digital foetal monitor for expectant mum's concept (see below).

Figure 7.5: Digital Foetal Monitor for Expectant Mums

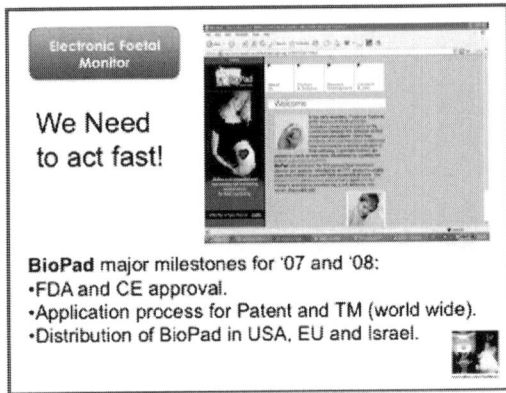

Source: UK Final Presentation, ISF Project.

The UK team submitted twenty-seven ideas and (Figure 7.6) all but one of them (or 96%), require or suggest new technology incorporating new scientific knowledge. To bring these ideas to the market would require that GSK commit to the acquisition of new specialist, scientific knowledge. Of the UK team's concepts, 50% were what could be described as radical –i.e., new technology in new markets. This reflects a very high level of novelty from this team, the ideas are totally unconstrained by the current competencies and technologies that are strengths of GSK.

By contrast, the US team's concepts were not skewed so much in favour of radical ideas. Of the seven concepts, three demanded new technology; two were intended for new brands or new markets and one qualifies as 'radical'. Most are in the area of current brands and markets for GSK; suggesting mainly line extensions for existing brands.

Figure 7.6: ISF Outputs; Radical V's Incremental

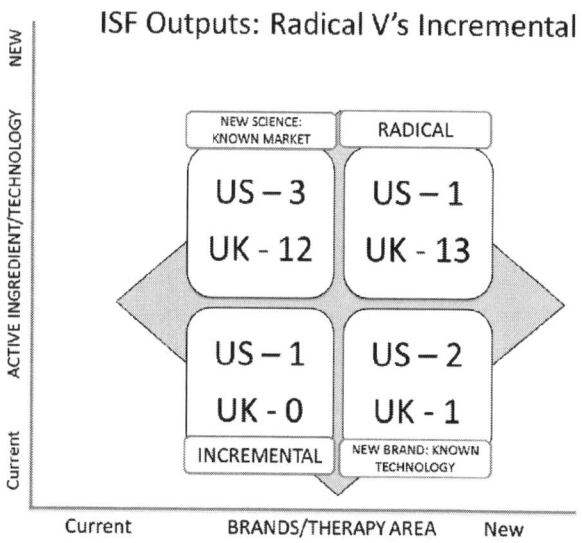

ISF Outputs: Radical V's Incremental

NEW

ACTIVE INGREDIENT/TECHNOLOGY

Current

NEW SCIENCE: KNOWN MARKET	RADICAL
US – 3 UK - 12	US – 1 UK - 13
US – 1 UK - 0	US – 2 UK - 1
INCREMENTAL	NEW BRAND: KNOWN TECHNOLOGY

Current   BRANDS/THERAPY AREA   New

Source: Analysis of the portfolio of ideas from both teams

This analysis delineates very clearly the variation in the outputs of the two teams competing in the ISF programme. From this analysis, I propose to separate the two teams, and to treat the UK team as an exemplar of Radical Innovation (because, this is what they delivered) and the US team as an exemplar of Incremental Innovation. This is not to elevate the UK team or denigrate the US; both teams delivered satisfactory outputs albeit with considerable variation between them.

Radical innovations have the potential to transform entire industries; incremental innovations provide low- risk potential for product upgrades (Hurmelinna-Laukkanen et al., 2008). Nevertheless, they contend, it is important for each firm to agree their own definition because not only do different types of innovation play a different role in the organisation's portfolio but they also require different management styles to facilitate them. In that context, the next chapter (Chapter Eight) will analyse the antecedents and factors which may have contributed to the development of radical ideas in the case of the UK team and incremental innovation in the case of the US. It will subsequently suggest an organisation-level of project-level framework which may be helpful for organisations who have a purposeful objective in pursuing either radical or incremental innovation or exploring both.

7.4 The Innovation Processes Employed in the Case Study

The following section will explore the processes used by each team in their response to the project challenge. Within this section, large themes like the idea generation approach is discussed. Idea generation is, in itself, a process, although it encompasses other processes within it. Moreover, it was the first phase in the project and the one in which the approach of both teams is contrasted.

In the case of the UK team, the bulk of their time, energy and effort was concentrated on the idea generation phase. Hence, it is proposed, for the purpose of clarity to include their use of both internal and external networks in this (idea-generation) section also. In keeping with the chronological narrative approach adopted in both Chapters Five and Six, it seems appropriate here also to access the literature through the richness of the teams' experiences in the sequence that they occurred. Hence, this section will begin with a review and discussion of the idea generation processes. It will then progress to processes used in the idea conversion phase, including the issue of idea prioritisation and voice of consumer (VOC) issues. It will discuss the practice and theory around innovation champions and their role in smoothing the path for ideas inside organisations.

7.4.1   Idea Generation and ISF – The UK Team Approach

The generation of novel and useful ideas is a necessary precursor to innovation. It is an essential ingredient for innovation projects. Amabile et al. (1996) propose:

> All innovation begins with creative ideas... We define innovation as the successful implementation of creative ideas within an organisation. In this view, creativity by individuals and teams is a starting point for innovation; the first is a necessary but not sufficient condition for the second.

The ISF project began with the idea generation phase. Both teams started with a blank page and both teams approached it differently.   The UK team embarked on a more substantial outreach or boundary spanning programme with research institutes and universities. Specifically, they made a site visit to the Surrey Sleep Research Centre in Guildford University, Surrey. Here they met and talked with the Principal Investigator on a number of sleep studies and these discussions formed some of the basis for two specific ideas that the team ultimately presented to the SLT.

The team leader also arranged a meeting with Bufton Consulting, an Open Innovation intermediary consultancy, located in Reading. Bufton Consulting were given a brief, consistent with the team's own brief, to go scouting for interesting technology in certain therapy areas and to report back to the UK ISF team with their findings. From this meeting, Bufton returned with some promising technology ideas in digital disease management. This is an area of healthcare which has subsequently been very well developed and is referred to under the heading of 'tele-health' or 'connected healthcare'. Bufton Consulting approached the UK team with ideas about digital diabetes testing and monitoring devices as well as foetal monitoring devices. Both these ideas led the ISF team into contact with experts in those fields and these ideas were developed during the project and were ultimately selected for inclusion in the final presentation.

The UK team leader encouraged the team to look for areas in which they had an interest or where there might be interesting, relevant new technology and to go and talk to external experts in different fields:

One of the things that we tried to set out to do was not to have anything too process-driven, but to provide people with the opportunity to learn about new areas, find out new information, speak to new people, from which they could generate new threads. So the philosophy was that rather than have a process whereby you go through different stats (market statistics) to try and pull ideas together, is you create an environment where people have the opportunity to read about new areas, talk to experts in different areas, interact with different people such that they can generate threads for their own development.

The UK team leader discussed with his team the two possible approaches, he saw, to their task. He contrasted the customer-driven, insight-led innovation with technology-push approaches. He concluded that his team should really focus on the technology route rather than the consumer insight. He had legitimate reasons for this approach; first his team was comprised entirely of technical people, scientists, medics and regulatory specialists and, following from that, they did not have any marketing or market research people or expertise on the team. In his interview, he describes his selection of this approach very clearly:

Now I would say there are broadly two areas where you can come up with huge wins in terms of new products. One is if you come up with a technical solution that is slightly different and can be positioned to an unmet or inadequately met consumer need. And that is what I would describe as a technically client-driven solution. Or the other way you can do it is if you just happen to strike on an insight, or a piece of consumer understanding, which enables you to make something of what you have

already got   available in a light that hasn't previously been recognised.
As a group of scientists I felt our role was very much to come up with new
technical solutions. Things that would at least be worth exploring from a
technical point of view.

Although not labelling it as such, Grist is actually describing the difference between technology-push and demand-pull innovation. He is declaring his preference for the former. The leader's intention to focus exclusively on the science struck a chord with other members of the team. One member, in particular, makes the point that GSK, in his view, has often had strong or promising technical ideas for which the company was unable to match an equally compelling consumer insight.

And I think as an organisation we have had that, many, many times in the past where we sometimes are not able to unlock the concept with the consumer straight away and we tend to back away from it and say oh it failed, it didn't score well and we don't keep going at it enough to understand why it failed and how we might change it to make it work. And I think there is lots of scientific opportunities that weren't understood. So overall I was extremely enthusiastic about the approach of looking for innovative new business opportunities, addressing it predominantly from a scientific, R&D perspective.

The UK team also prioritised the creation of external networks with thought-leaders in certain therapy areas which were outside the GSK current served markets. As already noted, they spent some time in the national sleep laboratory, talking to experts in the area of sleep and this input is evident in some of the ideas they ultimately presented. The top two ideas they presented centred on the sleep category. Following the feedback from Bufton Consulting, the UK team also looked very closely at the area of tele-health or electronic devices to aid self diagnosis or treatment. GSK is a member of an Open Innovation alliance called the Continua Alliance. In their own words, Continua Alliance is:

A non-profit, open industry alliance of the finest healthcare and technology companies in the world joining together in collaboration to improve the quality of personal healthcare.

In simple terms, the Continua Alliance connects healthcare companies whose NPD ideas may be leading them closer to medical device or electronic solutions with technology companies whose own NPD efforts might be increasingly leading them towards health and wellness products. The Continua Alliance works on big themes (or disease states) and invites suitable partners to collaborate. The themes include diabetes, cardiac disease and COPD (chronic obstructive pulmonary disease). The UK team went and met with people from Continua and discussed some ideas with them and, following that, they sought out and met with internal GSK experts on diabetes. In the Pharma division of GSK (by far the largest part of the company), diabetes is one of the company's specialist areas. When company personnel forge connections with subject matter experts in other parts of their own organisation, this practice is, and these relationships are often referred to as *bridging ties*. These are knowledge-sharing ties spanning internal boundaries in a formal organisation. Existing empirical studies demonstrate a correlation between bridging ties and improved innovation performance (Tortoriello and Krackhardt, 2010).

By connecting both with Bufton Consulting, the Continua Alliance (*boundary spanning*) and the GSK diabetes team (*bridging ties*), the UK team came up with a number of ideas around tele-health both in diabetes and in foetal monitoring during pregnancy. These ideas qualify as radical under our prior headings as they imply both a new brand from GSK, as they do not fit naturally under current brands within the company's brand portfolio, and they are based on technology, which is not currently being used within the company.

7.4.2 How the US Team Managed the Idea Generation Process

In contrast, the US team very quickly mechanised the process of idea generation. Team members were given dictaphones to record any ideas they had while out of the office. They were also provided with access and training on innovation software tool (ThinkTank® from Group Systems) with which they could list, cluster and prioritise their ideas online. Further, they were given a future trends presentation by a specialist trends company, called *Iconoculture*. This presentation made predictions about possible future trends in healthcare, wellness and self-medication and these predictions acted as prompts and stimulus for possible ideas for the team. In this way, the US team was making limited use of external consultants as part of their idea sourcing. However, they did not engage with suppliers or with universities and research institutes and this may have left a gap in their preparation. The team leader acknowledges:

> *As far as connecting with Research Institutes or KOL's (Key Opinion Leaders)... we didn't do a lot of this...and yes, most of our research was desk/literature/patent research-based. In just one instance, we did, however reach out to a world-renowned expert on medical applications for ozone... which served as input to a number of ozone-based new product/device ideas that were formally tested via concept study with BuzzBack.*

The US team leader intended to fully comply with the brief by covering the full spectrum of activities required, and so when he created his project plan, he allocated only the first nine weeks of the nine month project (25% of the project duration) to the idea generation phase (see Chapter Five). During this phase, the US team's approach to the idea generation task was to rely primarily on electronic brainstorming which is the use of computers to log, post, build upon and rank ideas. The team very rapidly developed a way of working that revolved around twice-weekly meetings. The brainstorming (or search for new ideas) was often done at these meetings but using the ThinkTank tool. Coapman, the team-leader, was a trained facilitator with this tool.

Software such as ThinkTank facilitates brainstorming without the need for face-to-face interaction. It also allows participants to contribute their ideas anonymously (Diener, 1979; Thompson, 2003). It allows team members to generate a lot of raw ideas without having to leave their own desks. Participants can also review the other participants' ideas and build upon them. This method of brainstorming is especially useful if team members are located on different sites where face-to-face meetings are difficult to arrange. It is paradoxical that it was the US team who embraced this technology when they were all co-located while the UK team did not use it despite being dispersed over two sites.

The US team were, when compared to the UK approach, very inward looking in how they scouted for new ideas. A number of the SLT members criticised the US ideas for lacking originality and being too predictable. One explanation for this is provided by one of the US team members, when she says that most of the US ideas were ideas that had already been floating around the organisation for some time beforehand and had become 'hobby-horses' for some of the team members.

> *Generating 300 ideas is very easy actually. Basically, we just find old ideas and put them in.*

There was one further, possibly subliminal, brake applied to the US idea generation process which came in the form of this same team-member's refusal to develop or even support any ideas that were, in her view, radical. She felt, that despite the company's stated desire to receive radical ideas for its pipeline, that GSK would never accept anything really novel or risky. This made sure that the ideas that she contributed were centred on modest performance enhancements of current offerings but she also discouraged her team mates from developing or promoting any ambitious, radical ideas.

> *'I do not think the company. I mean, even though we were told there's a*

no limit and the sky's the limit but ... I just do not think that the company were to go for it. Doesn't matter how much data, I just do not think the company was open to really new ideas.

In summary, the US team did not create an environment conducive to developing radical innovation even at the preliminary stage. By not developing or encouraging external, or even internal, networks, they were limiting the type and direction of the ideas that would emerge. On top of that, there were people on the team who considered that it was their role to protect the company against innovation and to actively discourage imaginative, novel, possibly radical ideas. However, the US team emerged from this phase and had, within nine weeks, assembled a portfolio of 300 ideas which they then went on to test.

7.5   Idea Conversion in the ISF Project

Variously referred to as *conversion* (Hansen and Birkinshaw, 2007), *transformation* (Roper et al, 2008) or *incubation* (Paulus and Huie Chuan, 2000); this phase involves selecting the ideas that have most appeal. As Hansen and Birkenshaw (2007) note:

> Generating lots of good ideas is one thing; how you handle (or mishandle) them once you have them is another matter entirely. New concepts won't prosper without strong screening ... mechanisms. (p124)

In the ISF project, this second phase, in the strictest sense, was really only undertaken by the US team. In the UK, the phase of idea generation continued right up to the end. Although, certain team members who had formalised their ideas did have conventional concepts ready for testing but for the most part, the UK ideas were very early stage, embryonic and not fully formed. By contrast, the US team really put time and effort into doing the second step to a high level. So, once a quantity of novel, high potential ideas are identified and collected, the next phase of the project involves what are often prickly decisions about which ideas to prioritise and which to abandon. Ideas often take on a very personal dimension, as each team member is naturally and especially fond of their own contributions and may be reluctant to discard their own ideas and embrace the ideas of others. This is an area where team conflict is almost inevitable as was demonstrated in this case study.

Once the selection decisions were made and the uncertainty, which characterises innovation, was overcome, then the US team entered a period of high performance. They managed the consumer research process, moving from qualitative to quantitative research very efficiently. They rendered their ideas very professionally, using expert copywriters and illustrators to convey their ideas in the most appropriate way for the research process. The project management skills of the team leader came to the fore as soon as they had left the 'fuzzy front end' behind them.

7.6   The Variation in Outcomes for the Two Teams

The large variation in approach and outcomes in the case study data prompts a number of questions; including why did the teams produce different types of ideas, and, more importantly, is the variation in process and leadership, observed in the previous chapter, associated with different outcomes. It could be argued that the UK team invested all their time and energy in the idea generation phase of the project; they concentrated on invention and never got around to either of the other two following phases; conversion and diffusion (Hansen and Birkinshaw, 2007). In that sense, one could suggest that the UK ISF team were not actually an innovation team as they were not active across the full spectrum of activities that constitute the innovation value chain. They confined their activities principally to the idea generation stage. Despite this (or possibly because of it), the UK team were credited with producing the more original, novel and radical ideas of the two groups and this may, in part, be attributable to them spending so much time in the creative realm of idea generation. It could also be connected to the fact that they built better internal and external networks to access knowledge, ideas, expertise and insight.

There were considerable differences in the teams' approaches to the challenge. The UK team, for instance, did not engage, in any significant way, with consumers. They believed that their mandate was to maximise the technical, scientific input into their ideas and this higher focus on the science is likely to have been a contributory factor to their ideas (being widely reported as) having more of a breakthrough feel to them than the US ones. The absence of customer feedback combined with their alliances with technology consortia (Continua Alliance) had the effect of making the UK ideas more technology-led than consumer-driven. Further, it could be argued that the UK team were deliberately pursuing radical ideas whereas the US team, with their very high level of consumer interaction and tight project management, may have been destined to, or certainly more likely to, develop incremental ideas. This interpretation will be explored further in the coming pages.

The more tightly project-managed US group coached all their members through what appeared to be a very positive developmental experience, for each individual involved, of running a consumer focus group. Each member of the team had to run a focus group with consumers from the target group for which their idea was intended. For many, this was the first time that they had ever interacted with customer research; and, in this case, they were not just attending it but they were actually moderating the group. In this context, it could be suggested that the US team were building an innovation capability within their team by building skills that would endure beyond the lifetime of the ISF project. On the other hand, the UK team were exclusively building ideas and not, intentionally, building capability.

The US group also managed to carry out both qualitative (as noted above) and quantitative research on their ideas which demonstrated that they performed better in the latter stages of the innovation value chain: conversion and diffusion. It's worth noting that these latter stages of the innovation process were part of the original brief and in carrying them our as part of the project, the US team were being entirely compliant. This compliance also points to a perspective that the US were more process focussed while the UK were more focussed on the outcome. One of the SLT members notes:

The US team's ideas were more line extensions for existing brands than genuinely original, novel, radical ideas. The US stuff was predictable and really just incremental.

If the teams could be plotted along a continuum stretching from *explore to exploit*, the US team would have been at the explore end with the UK closer to exploit. The US ideas almost all exploited existing brands and technologies whereas the UK team's ideas were predominately exploring new technologies and/or new markets. There is a difference between the 'R' and the 'D' in 'R&D' (Bain et al., 2001). Research is the phase where new ideas are very valuable while development is more concerned with bringing the new ideas into being. Hence, the UK could also be classified as a research Team with the US more of a development team.

One further classification that might separate the two teams is their approach to the challenge. The US team, with their adoption of project management tools, critical path, schedules and principles can be described as having followed a linear, stage-gate approach to the innovation process. The UK, whose approach was not formally managed through a sequence of stages or gates, adopted more of a network approach. The network approach, in their case, implies that they were more outward facing and adopted an approach more consistent with open innovation.

Table 7.3 below summarises some of the major dimensions upon which there was a significant difference between the approaches and processes of the two teams which seem likely to have had an inpact on the outcomes they ultimately produced. These issues are further discussed in the following two chapters.

Table 7.3: Approaches of the Teams to the Process of Innovation

Radical Innovation Based on UK Team	Factor	Incremental Innovation Based on US Team
Loose and flexible	Leadership	Tight and structured
Internal experts are sought out and their expertise is harnessed in building the concepts.	Internal Networks/ Bridging Ties	Internal experts on various promising therapy areas are not sought out or tapped into. The team is self-reliant.
The innovation team takes responsibility and does not actively build internal networks to support the ideas. They prefer to 'surprise' the senior managers.	Finding Champions for the ideas within Senior Management	The search for future supporters for their ideas is purposeful and actively managed. SLT are sensitised to what's coming; 'no surprises' is the approach.
Team is comprised of a group of individuals with little formal contact who pursue specific ideas through largely personal motivation. Team members largely unaware of each other's projects	Team working	Effort is made to channel the team's work, meet regularly, achieve consensus and to reduce team conflict. Team continually reviewing each other's work as a group
No formal project plan. Only one deadline.	Project Management	Plan developed with milestones, deliverables and critical path with a series of interim deadlines
Ideas are actively scouted from outside the company and relationships are initiated with external experts. Team is externally focused	External Networks	Experts asked to advise once only on a consultancy basis on projects that have been initiated internally. Team is internally focused
Consumer research is considered once the scientific basis of the ideas have been established. It has low priority. Technology-led	Voice of the Consumer	Consumer research dominates the process and dictates the type and level of novelty in innovation. It has top priority. Customer-led

Chapter Eight

Discussion

8.1   Chapter Objectives and Contents

This study began by asking what insights might be gained by studying closely the innovation processes and outcomes of a large, global and successful company. Specifically, the study explores how the company organises to encourage radical innovation. It emerges from this study of GSK that within a team-based innovation experiment, two sets of NPD ideas were developed; with one set qualifying as radical ideas for innovation while the other set was more incremental. This significant difference leads to the further question of what were the antecedent processes and issues that led to such wide variation in outcomes.

This discussion chapter has two objectives: first, to propose a framework which fuses the practice-based insights (derived from this study) with relevant theory with the objective of developing possible future guidelines for managing innovation projects in GSK and other large, R&D-intensive environments, and, secondly, to determine whether extant litertaure can furnish an adequate explanation for the variation in outcomes from these two teams.

In terms of extant literature, the case study is an example of a team-based innovation initiative that focuses on the earlier phases of the innovation value chain (IVC). Specifically, the case data presents variation in terms of types of innovation; radical and incremental. It also highlights variation across the innovation process, including variation in how the teams were managed. The case analysis identifies a number of themes that emerge from the study, which emerged through the analysis and coding process, reported in Chapter Six. These issues are as listed in Table 8.1:

Table 8.1: Key Themes from ISF in Managing Innovation Projects

Themes	Constituent elements
Structure	Making time for innovation
	Squaring-off line managers
	Size of team
	Space to innovate

Process	Ways of working (use of IT, co-location)
	Harnessing the voice of the customer
	Objective methods for idea prioritisation
Networks	Bridging Ties (internal)
	Boundary Spanning (external)
	Recruiting innovation champions
Leadership	Technical Skills
	People Skills
	Project Management Skills

The case described in this study is positioned in the context of (1) the activities occurred upstream in the innovation process; at the fuzzy-front end; the idea generation and conversion phases (2) the case is an experiment in team-based innovation and, specifically, (3) the case highlights considerable differences in the leadership of such teams. Although often graphically represented as a sequential flow, the innovation process is regularly described as non-linear (Caldwell and O'Reilly, 2003). The structure of this chapter follows the path taken by the project itself. It begins by looking at issues surrounding project initiating structures. Next, the idea generation phase, as it was managed by the two teams, is discussed and analysed. Other processes are then discussed with reference to the relevant literature; managing and prioritising the ideas, managing the voice of the customer, using internal networks and championing ideas within the organisation. Finally, the issue of leadership is discussed.

The figure below (Figure 8.1) graphically maps the stages of the IVC in which the two teams were active during this project. Their endeavours will be described in detail later in the chapter but this graphic illustrates that the UK team were really only active within the idea generation phase while the US team extended their efforts across all three although they stopped considerably short of getting a new product or idea launched or introduced.

Figure 8.1: Mapping the Teams' Activities Across the IVC

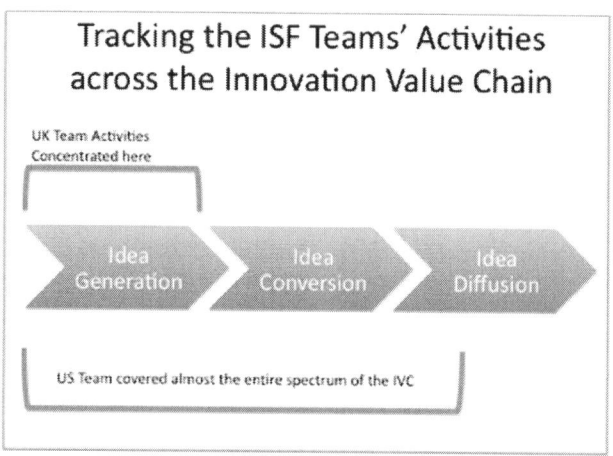

Tracking the ISF Teams' Activities
across the Innovation Value Chain

UK Team Activities
Concentrated here

Idea
Generation

Idea
Conversion

Idea
Diffusion

US Team covered almost the entire spectrum of the IVC

(Adapted from Hansen and Birkinshaw, 2007)

This chapter will now review the issues and themes emerging from the study
data. The first part will reprise the key issues or themes that emerged from the
case. Then, in the following section, each theme or tension point is discussed
in terms of how it is currently positioned within extant literature. The discussion
of each issue will conclude with some interpretation of possible implications of
the case for current thinking.

8.2   Overview of the themes from the case

It is widely acknowledged that, in terms of new product (and service)
development and their impact on both the industry and firm level, significant
differences exist between radical and incremental innovations (Dosi, 1982;
Christensen and Rosenbloom, 1995; Christensen, 2000). There a plethora of
classifications to define radical innovations (in particular) with some
emphasising their impact on the technology; others stressing their disruption to

the market; still others focusing on their novelty and more relating the radicalness to its impact on the parent organisation (See Chapter Two).

But many authors concur that firms must determine for themselves which of their innovations are deemed to be radical and which are incremental for two reasons; first, because firms need to balance their innovation portfolio and second, differing managerial practices favour different outcomes and firms need to organise accordingly. In this case, the challenge put to the two teams was to develop 'radical innovation' and for the purposes of this analysis, 'radicalness' is being defined as including ideas which suggest both new technology and new markets.

This chapter has followed the operating model of the Innovation Value Chain (IVC) (Hansen and Birkenshaw, 2007) which divides the innovation process, regardless of whether the outcomes are radical or incremental, into three discrete phases. Having examined the performance of the two teams, within each of the IVC stages, it is clear that the UK team elected to immerse themselves in phase one for the entire duration of the project and didn't really perform meaningfully against the 'conversion' or 'diffusion' phase.

Conversely, the US team spent only a quarter of their time in the idea generation phase after which they focused their attention on performing well in the two latter phases. Within both the idea generation and conversion phases, the US team neglected to build networks within the company (bridging ties) and outside the organisation (boundary spanning), confining their search to desk-based research and this omission seemed to limit the novelty or originality of the ideas they eventually presented. The UK team did, however, manage to include the sub-phases proposed by the IVC model; in-house, cross-pollination and external idea generation. It is suggested that the superior novelty of the UK team's ideas is due, certainly in part, to their external focus in the quest for original ideas.

Making this distinction between the two outcomes allows for an analysis of what precisely, in this case study, were the antecedents that led to the variation in outcomes. Specifically, in practice, what type of management would seem to better facilitate each type of innovation? Many companies will seek a balanced portfolio of new ideas with both radical and incremental opportunities and this analysis adds useful insights into the factors that favour one or the other.

Leifer et al. (2002) make the point that large companies have great difficulty in bringing radical innovations to the market:

> While it is clear that radical innovation is important to firms concerned with long-run growth and renewal, it is also clear that large, established firms have difficulty managing the radical-innovation process. Large, established firms have grown excellent at managing operational efficiencies, and at introducing next- generation products. However, the chaos and uncertainty that come with commercializing new technologies for markets that may not yet exist require vastly different competencies. (p. 102)

In this context, it is to be anticipated that there will be a higher level of managerial interest in the factors that promoted radical innovation within this case study than those that facilitated incremental innovation. One conclusion of this case study is likely to be that organisations can tailor their management practices to purposefully facilitate the type of innovation they seek. They can structure programmes and teams in a way, which is likely to lead to one or other, incremental or radical innovation, outcomes. Indeed, they can structure individual projects to match desired outcomes. As noted in Chapter Two, many organisations will actively manage quotas of projects in each area. Hence, these guidelines will be of significant managerial interest to managers tasked with managing and balancing an innovation pipeline.

The figure below (figure 8.2) summarises what are the principal findings of this case study. The first, as noted above is that there is a difference between radical an incremental innovation and that organisations can develop structures and processes and can choose team leaders in a way that will maximise their chances of developing either form of innovation. Analysing the individual structures and processes; conclusions can be drawn on the impact of various elements on programme outcomes. Four principal headings emerge from this case study analysis: structure, process, networks and leadership. The figure below shows how these issues co-exist and overlap in innovation projects.

Figure 8.2: A Four Factor Framework of Innovation Based on ISF Programme

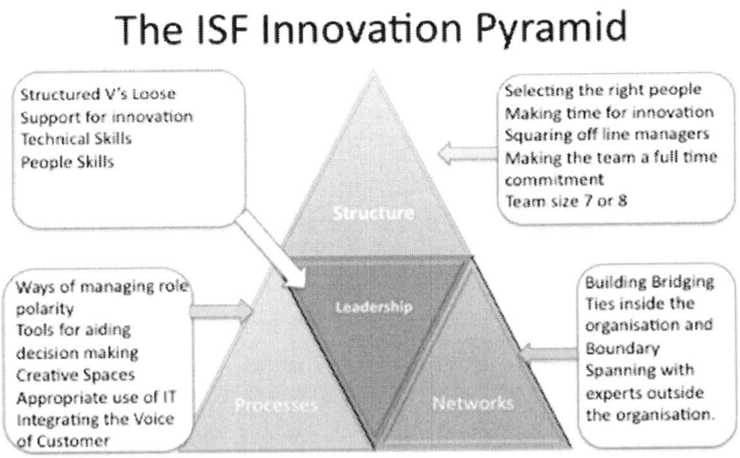

The ISF Innovation Pyramid

Structured V's Loose
Support for innovation
Technical Skills
People Skills

Selecting the right people
Making time for innovation
Squaring off line managers
Making the team a full time commitment
Team size 7 or 8

Structure

Ways of managing role polarity
Tools for aiding decision making
Creative Spaces
Appropriate use of IT
Integrating the Voice of Customer

Leadership

Building Bridging Ties inside the organisation and Boundary Spanning with experts outside the organisation.

Processes

Networks

Source: Graphic developed from case study findings

8.3   A Four Factor Framework for Practice

The table above is based upon the insights from the practical case of the ISF. This is a possible framework identifying the principal areas of tension that might affect the outputs of innovation teams in comparable settings. This representation reflects the layers of issues that need to be taken into account when embarking on a team-led innovation initiative or activity. It acknowledges the elementary (but nonetheless important) structural issues around establishing the ways of working and selection of team members.

Initiating Structures

These issues connected with the initiating structures of the project were shared by both teams insofar as both had to cope with time constraints; both teams had to cope with finding a space to innovate; both teams had to deal with the issue of providing cover for team members to work on the ISF project.

Although, these were reported as significant constraints in the experiences of the team members; they did not appear to materially affect the outcomes produced by the teams. In other words, both teams produced satisfactory outcomes despite the constraints of time shortage and the need to balance other work commitments. There are four elements to how initiating structures impacted on this case study. First, on a practical level, the notional assignment of 20% of people's time just didn't seem to work for the individuals involved. Many reflected that a more intense commitment allocating more, or preferably, all of their time for a shorter duration might have resulted in superior outcomes, although this is impossible to test. A second element that was reported to have had a negative impact on the project was that the SLT did not adjust anyone's workload, or make any effort to adjust it to take account of participants' involvement in this project. This seems somewhat naïve in hindsight; to expect the teams to deliver radical breakthrough ideas for the organisation but to do it without interrupting their other activities.

Compounding this, the SLT neglected to clear the participation of certain team members with their immediate line managers. This resulted in the line managers, being reported as, becoming a barrier to engagement with the project. Such a reaction could easily have been anticipated and dealt with. It was always possible that the line managers would feel overlooked and might possibly, as a consequence, try to thwart the input of their subordinates. Failing to legislate for this in planning the project, meant that participants had to almost conceal their involvement on ISF from their managers; or certainly be overly discreet about it and their contribution to it became seen to be almost subversive.

The third element within the initiating structure was that those members of the team who had most line responsibility, or, specifically, most people reporting to them found it most difficult to sustain a high contribution to the project because of the competing demands on their time. In the UK team, this was most acute. Some of the senior managers perceived that there was little structure to the project and little direction coming from the team leader and so some, stealthily, started to withdraw. UK team members who were located on a separate site from the main group were particularly affected. Ultimately, this group all pulled out of the project. But the withdrawal of some of the senior managers had one positive consequence; it gave the younger, junior team members the opportunity to take a more central role in the project.

Nevertheless, one would have to assume that the project could have benefitted from the wholehearted commitment and engagement of the entire group rather than the reluctant and fleeting connection of some (usually senior) people and the enthusiastic energy of others (usually junior). What happened in the UK team was a function of both, initiating structures where no provision was made to lighten the workload for the members during the period of their involvement, and of the leadership of the team which failed to provide clarity in directing the team members' work.

The fourth element was simply the size of the team. Reflecting on the experience, the universal view from the participants was that the teams were too large. Starting with 12 or 13 members made coordination of meetings very difficult; it made decision making difficult and it allowed certain people to hang back, even withdraw and not contribute as much as others.

Space to Innovate

At a prosaic level, one of the elements that made a positive impact, according to both teams, was when they were assigned their dedicated innovation team room. It seems to make intuitive sense that if you want people to break out of the routine and to think differently, then they are more likely to do it in a place which is away from their own routine; away from their desk. Assigning a dedicated space also allows them to store and exhibit stimulating samples, ideas, images and other team related artefacts in one place where people can go and interact with them. Having a team room or a project room provided a physical space to go for meetings and for individual work. It also seemed to provide a psychological boost in demonstrating the importance of the initiative by assigning it a dedicated space.

Other elements of the initiative appeared to be more highly associated with successful innovation outcomes. Elements such as how the teams were managed and how well connected they were, both inside and outside the organisation, (especially the latter) had far more influence in delivering superior outcomes. Hence, issues of initiating structure are hygiene factors; implying that while they need to be managed properly, they will not, in themselves, deliver the desired outcomes. Additionally, even if they are poorly managed, as they were in this case, such factors, alone, did not seem to stifle innovation to the extent that they might derail the entire project. Consequently, in the diagram above, they appear on the bottom of the table.

Prioritising Ideas

Prudent managers should also anticipate handling issues around managing the selection, screening and ranking of ideas in a way that does not cause unnecessary team conflict. Ideas should be screened in a way also that minimises the team's chances of selecting an inappropriate idea or, conversely overlooking a potential winner. One possible solution to resolve this would have been to have a commercial person on the team. In that way, the R&D specialists could concentrate on the science and technology while the commercial person investigated the likely market application and sales potential of the ideas being worked on. There was a concern, especially in the UK team, that minor ideas were receiving the same attention and resource from the team as potentially breakthrough ideas but that they were not equipped or sufficiently experienced to be able to tell the difference. Of course, there is no foolproof guideline for establishing the future value of ideas and the more radical the idea is, the more difficult it is to assign a value to its commercialisation. However, within GSK, there are a team of specialist commercial people whose role it is to make these type of, albeit necessarily fallible, projections. Such a specialist would have been helpful on the team.

In the absence of a commercial person to help with these prioritisation decisions, a bespoke (decision) framework would have been a helpful way to rank the ideas. But without such a framework, the UK team were susceptible to a disproportionate influence of personalities. They readily conceded that many ideas made it through to the final rounds more on the strength of personality of the originator than on any demonstrable, inherent quality of the idea.

Examples abound of frameworks for ranking and prioritising ideas but naturally the better ones are customised to fit the specific context. Neither team developed their own framework, nor did they make use of a standard template for one. Not having one left their decision-making very open to influence but, on the positive side, it may also have allowed them the flexibility to include more radical ideas. Frameworks are also fallible and can often favour ideas that are most adjacent to the current repertoire. Hence, they can discriminate against radical ideas. Arguably, some framework would have been useful for the team but not one that was too rigid. One that gave direction for prioritisation would have been helpful but it would also need to have sufficient flexibility to allow for truly novel ideas.

Voice of Customer

The four-factor framework derived from ISF also suggests a balanced view of integrating the voice of the consumer/customer into the process and argues that how and when this is done can be critical to the overall project outcome. It also argues that the prominence given to the customer input can have an inverse relationship with the novelty of the ideas being progressed. Customers can, in some instances, have a narrow view of the market and may be less well disposed to ideas which, in their view, involve radical changes from the status-quo. One of the reasons that the US team was declared the winners of the competition was that they had managed to verify their ideas through consumer research. They had carried out two rounds of consumer research by the time the ideas were finally presented to the SLT. The ideas had performed well in research and therefore it was concluded that there was some proven consumer interest in the ideas. However, the dominance of the consumer research element of the US team's approach gave rise to considerable controversy with other members of the SLT who vehemently disagreed with the priority it was afforded in that group. They believed that the ambition and quality of the science was diminished by the focus on market research. Moreover, they believed that if you put a bunch of high-performing R&D people together, it seems a poor use of their time to use them in managing a marketing research project.

This debate brings to the fore the issue of how to harness the voice of the customer into innovation projects. The question, in the case of GSK, is not if but how. This is a delicate balance; managing to integrate the voice of the customer in the development of ideas in a way that strengthens those ideas rather than diminishes or constrains them. It seems that if the voice of the customer is too dominant, that the ideas will be biased towards merely incremental changes. But if the voice of the customer is absent from the process, there is a high risk that the project will be so divorced from consumer wants and needs that it will fail to be commercialiseable. Once again, like the innovation process itself, this is a contingent model with no single, simple correct way of doing things in all instances.

In the ISF project, both leaders took opposing views to the importance of the market research element of the project. In the US, the team elevated this element and devoted up to three quarters of their time to managing it, showing evidence of their market pull approach. While the UK team leader wanted the science to lead the process and his interaction to researching the appeal of his team's ideas was perfunctory. Thus the UK team exemplified a technology-push approach. It seems that both were extreme perspectives, where a middle course might have been more suitable.

Networks

The framework indicates the high value to the team of purposefully managing their network of professional contacts both within and outside the firm. This is seen as particularly impactful in the quest for radical innovation. Networks played a central role in this case in three crucial areas: external collaboration networks; internal expert networks; and internal networks of idea champions

Bufton Consulting, a commercial open innovation intermediary, were called in by the UK team leader and were provided with the same brief that the team themselves had received. Bringing them in and asking them to respond to the brief with ideas and contacts proved to be a decisive move for the UK team. Bufton identified both specific areas of interest as well as potential commercial partners that fed into the UK team's final presentation. It was, in some part, due to the Bufton-inspired ideas that the UK team were generally considered to have proposed the more original and novel ideas.

In contrast, the US team did not engage any external partners for the idea-generation stage of the process. This was one critical difference between the two teams. The US team's ideas were criticised for being unoriginal and predictable and this is not surprising as one of their team members suggests that many of the ideas they proposed were ideas that had been suggested many times already (See Chapter Five) and that the ISF, for some people, was merely the latest platform through which they could promote favourite ideas. But by bringing in external consultants at the early stage, the UK team helped ensure that their initial palette of ideas was fresh and original.

The UK team also performed well in making connections within GSK with subject matter experts in different parts of the company. Hence, internal experts on diabetes were consulted for input to some of the ideas being hatched in the UK team. This connection with internal experts strengthened the ideas and built on the knowledge already residing in the overall organisation. The US team confined themselves to desk research and did not make any external or internal connections during the idea generation phase.

However, where the US team did perform well was in actively seeking support for their ideas in advance of the presentation. The US team leader held numerous meetings with the individual SLT members to let them know about the type of ideas that his team were working on; he thus sensitised them to what was likely to be presented in the final meeting. So, when the US team presented, there were no surprises for the SLT, they each knew in advance what was being proposed and had been briefed in detail on any ideas that were specifically in their area of responsibility. Interviews with the SLT confirm that they preferred the US approach of keeping them abreast of developments within the project. Some expressed frustration at the UK's failure to 'check in' at any time during the nine months; even though such autonomy had been part of the initial project design.

Team Leadership

Finally, the framework illustrates that the selection of team leader is a crucial one. A strong team leader is likely to have the technical expertise to manage the task; the creative ability to deal with the ambiguity required; the soft people skills to effectively leverage the talent of the team and keep them motivated and focussed throughout the journey; the political acumen to position the ideas with senior management to increase their chances of getting traction in the pipeline; the commercial judgment to be able to make prioritisation decisions accurately and the experience to facilitate the process effectively. Effective project leadership has been identified as one of the most important mechanisms not only for managing team dynamics but also for steering the teams successfully and efficiently through the new product development process (McDonough and Griffin, 1997). Thamhain (2006) suggests that effective project team leaders are social architects who understand the interaction between organisational and behavioural variables and can act accordingly to satisfy both.

Consistent with prior studies, this case study reinforces the link with transaction leadership and incremental innovation and between transformational leadership with radical. The model below outlines, in summary, how managers can organise for the type of innovation they require to suit varying objectives within an organisation.

Table 8.2: Leading Teams for Radical or Incremental Innovation

Radical Innovation	Factor	Incremental Innovation

Based on UK Team		Based on US Team
Loose and flexible Transformational Contingent	Leadership	Tight and structured Transactional Systematic
Internal experts are sought out and their expertise is harnessed in building the concepts.	Internal Networks/ Bridging Ties	Internal experts on various promising therapy areas are not sought out or tapped into. The team is self-reliant.
The innovation team takes responsibility and does not actively build internal support networks for the ideas. They prefer to 'surprise' the senior managers.	Finding Champions for the ideas within Senior Management	The search for future supporters for their ideas is purposeful and actively managed. SLT are sensitised to what's coming; 'no surprises' is the approach.
Team is comprised of a group of individuals with little formal contact who pursue specific ideas through largely personal motivation. Team members largely unaware of each other's projects	Team working	Effort is made to channel the team's work, meet regularly, achieve consensus and to reduce team conflict. Team continually reviewing each other's work as a group

357

Radical Innovation Based on UK Team	Factor	Incremental Innovation Based on US Team
No formal project plan. Only one deadline.	Project Management	Plan developed with milestones, deliverables and critical path with a series of interim deadlines
Ideas are actively scouted from outside the company and relationships are initiated with external experts. Team is externally focused	External Networks	Experts asked to advise once only on a consultancy basis on projects that have been initiated internally. Team is internally focused
Consumer research is considered once the scientific basis of the ideas have been established. It has low priority.	Voice of the Consumer	Consumer research dominates the process and dictates the type and level of novelty in innovation. It has top priority.
Technology-led		Customer-led

This project also reinforces the approach of situational leadership by demonstrating that the ideal type of leadership for an innovation project is determined by the stage of its development. Within this case study, each team captain exemplified an extreme end of the leadership spectrum. The UK leader refused to allow any process interfere with the loose, flexible approach to facilitation that he favoured. By contrast, the US team leader did not allow any loose or flexible conditions to pertain as he managed the project with tight metrics and timelines. While both approaches delivered some level of success within the project, neither, from the point of view of the participants was ideal. The UK approach did little to foster cooperation between team members or to channel their work. While the US approach was very mechanistic and process-driven and was reported to have constrained the fluid conditions necessary for creativity.

8.4   Positioning the themes within the literature

This next section positions the issues that emerged from the GSK case study (alluded to above) within the appropriate literature in order to see if such literature provides a satisfactory explanation for the stark variation in outcomes from the two teams. This study suggests that a four-factor model (Figure 8.2) provides a project-level framework for managing radical innovation projects. In that context, it is worth reviewing prior contributions from research of the factors that have been found to be connected with the theory and practice of radical innovation.

Factors connected to radical innovation

Many of these factors found in prior research are systemic company or industry characteristics. These are factors, generally, outside the control of an individual project or team leader. Hence, this study is of particular interest to managers as it presents a practical framework which they can readily apply at the level of the project or team. .

Table 8.3: Factors Associated with Radical Innovation

Factors correlated with radical innovation	Key Authors
• Learning and continuous improvement culture	Bessant and Francis, 1997 McLaughlin et al, 2008
• Active User Involvement in NPD	Von Hippel, 1988
• Overlapping and parallel working	Wheelwright and Clark, 1992
• Appropriate structure for managing innovation projects	Cooper, 1984 Wheelwright and Clark, 1992
• Cross functional teams with flexible problem solving capability	Jassawalla and Sashittal, 1999; Sapsed et al, 2002
• Use of prototyping tools, techniques and design-thinking	Dodgson et al, 2005 Brown, 2008 Verganti et al, 2007
• Use of a radical innovation hub or centralisation	Ettlie et al, 1984 Leifer et al, 2007
• High levels of individual and organisational creativity	Amabile, 1998 Taggar, 2002
• Organisational tolerance for ambiguity, failure and risk	Nembhart, 2009 Kalunzy et al, 1972
• Strong and active market understanding and	O'Connor, 1998

insight; ability to interpret weak signals	
• Relationships, collaborations and networks	Bessant et al. 2003; Reed and Walsh 2002, Chesbrough, 2003 Birkinshaw et al, 2007
• Clear and active support from senior management	McDermott and O'Connor, 2002
• Lengthy development stages and associated high cost	McDermott and O'Connor, 2002 Golder et al, 2009
• Firm size and position in the market	Christensen, 2007 Chandy and Tellis, 2000
• Team Climate for Innovation	Anderson and West, 1998 Caldwell and O'Reilly, 2003
• Start-ups and new entrants behind most discontinuous innovation	Bessant et al, 2004

In terms of the issues which surfaced as signicant in the case, the following section probes the literature for guidance on how these issues should or could have been managed within the project.

8.4.1 Structuring for Innovation

As already mentioned in this section, the following issues emerged as significant in the context of initiating structures for innovation projects:

1. Devoting the right amount of time to innovation
2. Squaring off the participants' line managers
3. Getting the size of the innovation team right
4. Finding physical spaces conducive to innovation

These were significant issues for managers involved in the project. Storey and Salamander (2005) conclude that there is considerable value in attending to managers' experiences and insights because these same managers set the tone for everyday talking and thinking about innovation within their firms. Moreover, they offer important data for understanding which combination of factors act to promote innovation and, conversely, which combination stifles innovation.

8.4.2   Devoting the Right Time for Innovation

Best practice guidelines for product development teams usually take one of two possible directions; human resources practices or technical guidelines (Pattit and Wilemon, 2005). As this case-study demonstrates, both those approaches have a contribution to make but neither is, by itself, sufficient. The allocation of sufficient time to devote to a project is a hygiene factor in any organisation and the fact that this basic element of structure was absent (or, at best, poorly managed) from ISF speaks to the lack of effective, practical senior management support for the project. Swink (2000) highlights that senior management support for NPD practices, such as cross-functional NPD teams; enable team members to accomplish their goals by providing vision, direction, enthusiasm; but crucially, priority and access to required resources. This provision of access to resources and designation of ISF as a priority activity was missing in the execution phase of this project.

The literature on the effective management of R&D or innovation teams is silent on the issue of making time, clearing calendars, removing obstacles and allowing the team the time to devote to the task. Yet, this is a basic, necessary but insufficient condition for a climate of creativity to flourish. The identification of a low level issue like time availability, which could appear elementary and intuitive and yet is reported to have such a significant impact on the outcome and experience of the project, is one of the merits of the case study approach. Such a routine issue may not have been picked up through a survey instrument unless it was specifically identified in advance and explicitly probed in the fieldwork.

Similarly, in their study 'Views from the Trenches', Barczak and Wilemon (2003) found that the most frequently cited source of conflict impacting NPD teams revolved around similar 'company-level systems, policies and procedures' over which the team itself has limited control. 54% of their respondents cited conflict with senior management and other organisational units (functional departments) as the major sources of company-level conflict. Friction with senior management was concerned with policies and procedures, support, and resources. They note:

Specifically, respondents noted that conflict often resulted in negative feelings about the project, frustration, and stress. These feelings, in turn, often affected morale and commitment to the project. (Barczak and Wilemon, 2003: p 473)

Barczak and Wilemon (2003) also asked team members about the major sources of stress they experienced on NPD teams. 37% of their respondents cited 'schedule pressures' as a stressor in NPD. This was the single biggest 'source of stress' cited in the survey and the ISF experience confirms its significance.

The ISF (Innovation Sans Frontiers) began with the assumption that a commitment of twenty percent of participants' time would remain a fixed ratio throughout the duration of the project. This assumption ignores the phases of the innovation process and how practitioners may need the flexibility to get more or less engaged as both the nature of their specific input and the intensity of the overall project demands. In extreme circumstances, when the potential of the project warrants it, the organisation should build in flexibility for employees to devote more, or even all, of their time to certain promising projects (Boyle et al., 2003). It is further recommended that team members' reporting relationships should remain flexible so that, should it be necessary, reporting lines could be adjusted for the sake of the project. In this context, Hellinghausen and Myers (1998) suggest that 'functional managers represent one of the largest obstacles when implementing cross-functional teams, mainly because such managers will be required to give up some of their power so that the team can accomplish its goals. This was also the case in the ISF project as some line managers actively discouraged their direct reports from spending time on the project. Storey and Salaman (2005) also found that in poor-performing innovators, it was often the case that managers saw their role as protecting the firm from innovation, which in some cases, they saw as subversive and even childish. In such cases, managers themselves, represent one of the biggest barriers to innovation.

Additionally, Warwick University's KIN (Knowledge and Innovation Network's; 2006, p. 11) benchmarking study of innovation communities noted that Project Leaders were frequently asked to assume the responsibility for projects 'part-time' and would be expected to find the time required. The 'KIN' recommendation is that companies give their people adequate time to work on these projects rather than merely hope that their executives will 'find the time'. Govindarajan and Trimble (2010) argue that best practice in innovation projects suggests that each project has a dedicated resource (person or team) and this is augmented as necessary by technical or functional specialists from within the company. They conclude that organisations cannot simply ask and expect employees to participate in an innovation team where there is no full time, dedicated resource to manage the process or project.

This facet of the project design could have been anticipated by the SLT (Senior Leadership Team). Additionally, by neglecting to bring the functional line managers 'on-side' with the project, they left the team members open to a high level of potential conflict and personal stress. This conflict and stress was clearly reflected in the actual experiences of the team members. Moreover, there seems to be consensus from the participants that it would have been preferable to assign or second people to the project full-time, even if for a shorter period, than have this notional 20% of time dedicated to it over the longer period.

The case data highlights the importance of these issues to the experiences of the team members involved in the ISF initiative. While highlighted in selected studies, this is not an issue largely studied in the literature.

8.4.3 The Right Size for an Innovation Team

Research on team size indicates that teams are most effective when they have enough, but not more than enough, members to perform the group task (Guzzo, 1988; Guzzo and Shea, 1992; Hackman, 1990). Academic research by both Bouchard and Hare (1970) and Renzulli et al (1974) found that the output of creative ideas on a per-employee basis decreased as team size increased. Some research on group structure implies a curvilinear relationship between group size and innovation. Very small teams, of three people or less, it is suggested, will lack the diversity of viewpoints and perspectives necessary for innovation (Jackson, 1996), whereas large teams (above twelve or thirteen) will become too unwieldy to enable effective interaction, exchange, and participation (Poulton, 1995).

There is much support in the literature for the use of small project teams (e.g. Whitten, 1995; McConnell, 1996; Carmel and Bird, 1997; McConnell, 1998; Sawyer and Guinan, 1998; Humphrey, 2000; Jiang and Klein, 2000; Whitehead, 2001). As the size of the group gets larger, the difficulties of agreeing objectives, ensuring appropriate participation in decision making, achieving consensus on what constitutes high quality, and getting unanimous support for innovation, all increase (Curral et al., 2001). Specific details on precisely the optimum number to include in an innovation team can vary with industry and will depend on the complexity of the initiative. Foote's (2003) survey of new product development teams in the beverage industry showed:

Proving the old adage that too many cooks spoil the broth (or the brew), less than one quarter of companies report having more than ten members on their NPD team. Nine is the average team size. (p12).

Previous research and theory suggest that increasing group size will hinder effective innovation (Curral et al., 2001) since there will be physically more team members who need to reach agreement on actions and decisions; more who will seek to have their say in decision-making; and more who will interpret and discuss quality of task issues. Pozen (2010) claims that within groups of 10 or more, members engage in what he calls "social loafing": They stop taking personal responsibility for the group's actions and rely on others to take the initiative and get things done. Large groups also inhibit consensus building, which is a vital ingredient in innovation teams. The more members there are, the more difficult it can be to reach agreement, and consequently fewer decisions are taken and, consequently, less gets done

Tasks demanding high levels of innovation generally pose high demands on team members including remaining united in the face of external conflict and resistance, and managing the uncertainty necessarily connected with innovation. Curral et al. (2001) demonstrated that increasing team size has a deleterious effect on innovation-related group processes under task conditions that require high levels of innovation.

In summary, teams should be comprised of a sufficient number of members to have, between them, the technical knowledge required to accomplish the innovation task. But such teams should not have higher than sufficient numbers as this has been shown to have a negative impact on team processes and, ultimately, team innovation performance. The ISF teams were originally planned to have twelve members each and this was thought to be too many by the participants as they struggled to function effectively using all their members. The literature, too, supports the view that twelve is too many and that between six to ten (obviously, depending on the task) would be more advantageous in most circumstances.

While the data does not allow for an evaluation of what would represent the 'right' size for an innovation team, it does highlight that it may not be team size that matters most but rather team management and leadership. Both teams, nominally, were similar in size but the UK team had a significant number of members who did not engage. The process described in the UK, what the case suggests is the radical innovation team, is that it was 'individuals' and not the team that drove the development of ideas. So, in the context of the extant literature that focuses on the right sized team, this case data suggests that this question may be much less important when the task specifically mandates radical innovation.

8.4.4 The Importance of Creative Spaces for Innovation

Many practitioners, when providing guidelines for best practice in successful business creativity will point out the impact of the physical working environment on people's ability to be creative (Reinersten, 1997; Kelley and Littman, 2001). This advice seems especially pertinent to the early stage, fuzzy-front-end of the innovation process where novel ideas are being specifically sought and encouraged.

There is an intuitive logic to the idea that because innovation, by definition, implies exploring the unknown, that it should be facilitated by breaking the routine of the 'day job' and physically moving around into a different and more creative space. Collaboration Rooms, Creativity Labs and Innovation Centres are becoming increasingly popular in large organisations (Wycoff and Sneed, 1999). Most conventional meeting rooms are designed for presentations rather than collaboration. Case-based examples abound in the literature for example; Suzanne Merritt buying used furniture and prevailing on family and friends to paint vibrant colours on the walls of what became the Polaroid Innovation Laboratory (Merritt, 2010). Gordon MacKenzie of Hallmark is known for having created their innovation engine room by finding old roll-top desks for sale and using abandoned milk churns as waste paper baskets. As another example, the technology innovation firm, IDEO is often cited (Kelley and Littman, 2001) as an example to be followed in this regard for the fact that they allow their staff to bring their bicycles inside the office and hang them up on various pulleys located near workstations. One of their staff has an aircraft wing suspended from the ceiling over his desk and staff are encouraged to bring in 'cool new stuff' which is often showcased on a number of technology trolleys that people move around the building and use the contents as stimulus for brainstorming for various projects.

In the case study, while the original project design had not legislated for the teams to have any dedicated space; this was facilitated on request and it appeared to play a positive role for both teams in three ways:

a)   It enhanced the teams' sense of identity having a room exclusively dedicated to the business of the project

b)   It provided the teams with physical space in which to have their meetings and to store their prototypes and project paraphernalia

c)   It was especially useful for those managers whose own line managers were not supportive of the project by providing them with a private space to get out of the way and discharge their duties in relation to the project.

The case data supports a positive effect accruing to innovation efforts when management provide a dedicated space for the team to meet and collaborate together. In the case of the ISF project, both in the US and UK, once a dedicated room had been allocated to the team, it had an immediate and beneficial effect. The rooms in question were conventional meeting rooms in R&D buildings and could not be described as creative or zany in any way. Nevertheless, just having a dedicated space for the team to meet and work together appears to be conducive to both types of innovation.

8.4.5 Processes in Organising for Innovation

The case data highlights two key process differences between the teams that appear to be instrumental in the outcomes they ultimately delivered: first was the way they prioritised their ideas and second, was the way they harnessed the voice of the consumer into their ideation process. These two processes are closely linked. A third process, in which there was also considerable variation in the two teams' approach, was their use of specialist software and technology to assist with the process of idea generation. This third element did not appear to be a significant factor.

8.4.6 How to Prioritise Ideas and Manage Conflict?

The major factor identified in the literature that impedes creative performance is control (Amabile, 1998; Angle, 1989; Kanter, 1983; Oldham and Cummings, 1996). This is generally a controlling form of management where autonomy is restricted. A culture that supports and encourages control will result in diminished creativity and innovation (McLean, 2005). The primary reason for this is that too much control negatively affects intrinsic motivation. According to Amabile (1988), whose (KEYS) model of creativity asserts three equal components of organisational creativity: technical expertise, creativity and intrinsic motivation, asserts that expertise and creativity skills must be accompanied by intrinsic motivation to produce highly creative behaviour. The output of such creative behaviour, in the context of NPD, is a set of potential new product, service or business model ideas.

Once the ideas were generated, a next step (which followed really only in the case of the US) was to rank and prioritise the ideas which were likely to feature in the team's final presentation. Two key points are central to understanding this element of the theory and practice of team-based innovation management. First, there will inevitably be task conflict in the innovation process; the nature of the process renders that inevitable. Second, the conflict is likely to be most acute at the point of the process, which is most convergent. The innovation process is often represented as a series of steps with each one having an element which is divergent; where all ideas are welcome; where quantity takes priority over quality and where no judgment is involved. There is rarely conflict at this phase because no selection or prioritisation has to be made. But invariably every project arrives at a point where choices have to be made and where some ideas are eliminated and others progressed. Loewe and Chen (2007) represent it thus (Figure 8.3 below). At each phase, it is the convergent element, where certain ideas are being jettisoned, that will give rise to conflict.

Figure 8.3: Convergent and Divergent Phases of the Innovation Process

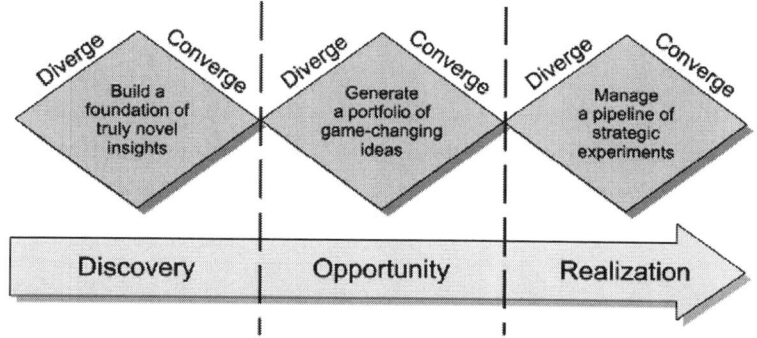

Ozer (2003) notes that two types of mistake are likely to occur at this point of the process; either the company passes over an idea which had genuine and significant potential or the company pursues an idea, which subsequently fails in the market. In either case, firms could accrue big losses. The former leads to missed investment opportunities and the latter to high costs with disappointing return on investment. Consequently, it is clear that it is in the best interest of firms to make accurate evaluations of NPD opportunities.

Because innovation is such a complex set of processes, to describe and enact it, scholars and practitioners evolved a stage-gate model that almost all companies now employ in some shape (Wheelwright and Clark, 1992; Cooper and Kleinschmidt, 1997). Acknowledging both the complexity of the process and its centrality to the survival of most organisations, companies started to structure the process in the 1960's. Tushman (1977), based on the previous work of Marquis (1969), describes these emerging structures and separates out different phases of the innovation process: idea generation, problem solving and implementation. These are now integrated into a more recent model (Hansen and Birkinshaw, 2007), the *Innovation Value Chain* in which the equivalent stages are; idea generation; (idea) conversion and (idea) diffusion.

Schmidt et al. (2009) contends that the new product development (NPD) processes that exist in most organisations consist of just two basic elements: activities and review points (See figure 8.4 below). Activities basically resolve issues and assemble data about the viability of successfully completing the project. In between the development activities are review points where the project is reviewed and a decision is made to either go on to the next stage of the process, abandon it, or hold it until more information is gathered and a better decision can be made.

Figure 8.4: Activities and Review Points in a Stage-Gate Innovation Model

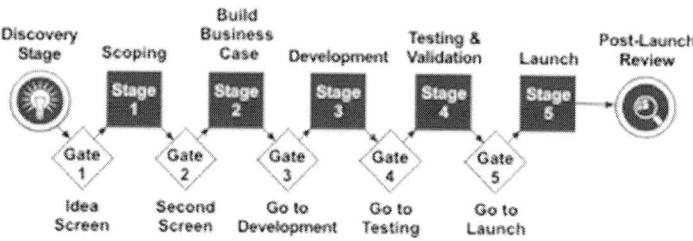

Source: Cooper, 1995

Hayes et al. (1988) created the original new product-development funnel that provides a framework for analyzing and thinking about how innovation unfolds in companies. This funnel concept is equally valid on levels of analysis for the organisation, the project or, as in this case, the level of the team. The first phase (Idea Generation) involves the creation of new ideas. At this start point, the main task is the creative widening of the funnel by knowledge acquisition, idea creation and exploration (Kratzer et al., 2006). For this phase, best practice suggests that ideas are not filtered or judged as the objective is to encourage the generation of as many ideas as possible (Kelley, 2001, p. 57).

Thereafter, the funnel must be narrowed by evaluating and prioritising the ideas and concepts, resulting in the selection of one, or a limited number, for prototype and possibly design for manufacture, followed by pilot production, manufacturing ramp-up, and, ultimately (if successful), market launch. This narrowing of the funnel, by necessity, requires the exit of a number of ideas as the process moves from divergent (i.e. all ideas are welcome) to convergent (i.e. certain criteria are being applied to the selection of ideas and some ideas will not survive the screening process). Cooper (1999) rates this as one of his top six success factors for successful product development: 'Tough go/kill decision points or gates—funnels, not tunnels.' It is this exit process where ideas are jettisoned by the group (resulting in the 'funnel not tunnel') that causes task conflict within the team as naturally individuals, especially the idea-originator, have become attached to certain ideas which may subsequently be earmarked for exit. Cooper (1999) noticed this phenomenon in his database and he described the inability or unwillingness of companies to kill projects as a major failing in corporate innovation management.

Howell et al. (2005) note the presence in some organisations of what they call 'stopping champions'. These are the opposite of 'innovation champions' who generally rally support to help ideas progress through the funnel. Conversely, *Stopping Champions* use political tactics and their networks to rally supporters to halt a project. They often achieve this by posing strategic and financial arguments for project withdrawal, and involving team members in establishing decision criteria to evaluate whether to continue or exit the project.

The stage-gates mentioned above have generally built-in decision criteria, which are applied to filter out non-performing ideas and concepts at all the major 'gates' of the process and they play some of the role of a 'stopping champion'. Edmondson and Nembhard (2009) hold that competing viewpoints, so essential for generating new ideas and making sound decisions can also lead naturally to conflicts that waste valuable time and can start to erode team relationships (as happened in both ISF teams). Indeed, as soon as a significant disagreement erupts in a product development team, dysfunctional behaviour and processes may be the rule rather than the exception, leading to frustration and flawed decisions (Edmondson et al, 2003; Edmondson and Smith, 2006). However, Clerq (2009) argues that the benefits of an innovation strategy for firm performance critically depend on the willingness of different functional departments (e.g., marketing, R&D) to accommodate conflicting viewpoints (De Luca and Atuahene-Gima, 2007) and this diversity, they argue, requires expert management to be harnessed for the benefit of the project.

Reflecting the combination of talents required to lead innovation in companies, Buijs (2007) titled his article: 'Innovation Leaders Should be Controlled Schizophrenics.' In it, he makes the point that team leaders, in order to deliver successful outcomes, have to manage conflicting agendas, objectives and personalities (p.208):

If the team is feeling down, then the leader should be optimistic; if the team is overly enthusiastic, then the leader should be cool. If the team has fallen in love with an extremely funny idea, then the leader should point out which were the original objectives of the innovative task. If the team rejects all of the ideas and they focus too much on feasibility, then the leader should provoke them to dream and to let at least some of the wild ideas be considered.

Disagreement arising from differences in opinions and perspectives of team members is labelled *task conflict* in the academic literature in contrast to other interpersonal conflicts possibly due to personality clashes and incompatibilities among team members (Jehn, 1995). Task conflict is referred to as *Team*

Polarity in the particular case of innovation teams and NPD projects (Van Engelen et al., 2001, Leenders et al., 2007). While team polarity may sound like a condition that warrants immediate repair and one that can only harm the team and its work, this is not necessarily the case. In fact, it is thought, contrary to the view expressed above by Edmondson et al, (2003), that team polarity can have positive as well as negative consequences (e.g. Souder, 1987; Van Engelen, et al., 2001).

On the positive side, team polarity, forces teams to consider perspectives and evaluate issues and opportunities more deeply than they otherwise might (De Dreu and Weingart, 2003). It can also make team members more flexible in their thinking (Carnevale and Probst, 1998). Conflict and disagreement about tasks can be helpful in defining the issues involved in the discussion (Putman, 1994); it can assist in developing new ideas and approaches to tasks (Baron, 1991). Indeed, frequent disagreement on alternatives has been shown to be a quality of successful teams (Bourgeois, 1980). Team Polarity is of most advantage to teams and initiatives when it occurs in the early phase when ideas are being generated.

However, the negative effects of team polarity can be harmful to teams. It has been shown to be detrimental to performance and satisfaction (Blake and Mouton, 1984; Pondy, 1967). Team polarity can impede the flow of work by delaying crucial decisions, especially if the team members are aware of each other's difference of opinion (Pelled, 1995). It can erode relationships; lead to flawed decisions and result in dysfunctional behaviours (Edmondson and Nembhard, 2009).

Therefore, a good team leader needs to have an ambivalence to team conflict because a certain amount of it is healthy but too much of it is dangerous.

However, the process for whittling down ideas from a high initial number to a more manageable group with some prospects of success, is not a straightforward one. The existing idea generation literature (often called the brainstorming literature), according to Girotra et al. (2003) is deficient in this regard because its focus is exclusively on the creation process and it pays less attention to the selection processes that teams need to develop and use to choose the most promising ideas.

Many companies have developed idea-screeners, which are decision aids (Ozer, 2003). In such a capacity, idea screeners should help decision makers make decisions more effectively and efficiently. However, if they complicate the decisions then they will defeat their purpose. In this (ISF) context though, a key criterion, based on the project mandate, for deciding to keep or abandon an idea was its sales potential. Sales potential will ultimately be a feature of demand and availability and these are impossible to know with precision or certainty at the idea generation stage. However, there are certain ways of framing broad market size and making assumptions either top down (i.e. what percentage of the market are we likely to capture) or bottom up (i.e. how many people will buy/use it; how often and extrapolate that for a national or representative population). Either way, these estimates are imperfect and provide a rough indication only of potential sales revenue. Hence, the type of conflict, lack of clarity and indecision described in the case study is inevitable unless the teams had agreed an a priori scoring system for the ideas.

Goffin and Mitchell (2005, p. 187) note that although financial considerations should be the dominant decision criteria for innovation projects, unfortunately the financial information available at this early stage is likely to be at best incomplete or else totally unavailable. This, they assert, is inevitable for a number of reasons; first, the completion date for the project may lie very far into the future and so there will be a high level of uncertainty about the condition of the market and the likely customer reaction in this context and this condition is more likely to prevail in the case of radical innovation. The second is that the preparation of a detailed financial plan requires a lot of effort that many companies find hard to justify when the idea is still an early stage concept.

Cooper and Kleinschmidt (1997) also found that even if detailed financial projections are made, innovation project managers often have little faith in them. The more successful companies in Cooper's studies are those that include financial projections as just one of a selection of more broadly based indicators. This, according to Goffin and Mitchell (2005, p. 203) is analogous to the use of tools like Balanced Scorecard to measure company performance. They argue that when financial measures are of questionable validity, the project selection process can be improved by including other criteria that have been shown to be correlated with successful new product development. Had they developed a scale or scoring mechanism of this type, the team polarity issue is likely to have been mitigated significantly.

This concept is broadly supported by the ISF story because, although the US team won the competition, it was generally considered that their ideas were the 'more predictable' or imitative while the UK ideas were seen as more original or innovative. So, although the US team was able to generate 300 raw ideas, in the end, the final seven they presented were not thought to be radical or disruptive to a high degree. However, in the case of the UK team, there were no such decision criteria nor were there any milestones or gates. This left the task of culling some of the ideas with lower potential up to individuals; a scenario which was inevitably going to lead to a high level of role polarity.

In the ISF story, it could be argued that neither team struck the appropriate balance; the US team had little if any team conflict and the UK, possibly, had too much. The ISF story reveals that both teams managed the first element of the process concerned with the generation of new, raw ideas successfully. Both teams report that they were able to generate in the region of three hundred raw ideas. The literature on creativity in the innovation process would suggest that the US team, because of the way it was (relatively) tightly managed, would have found it harder to generate creative, radical, new ideas.

In terms of implications for practice, the ISF story marks an interesting counterpoint to the theory. In theory, there should be task conflict at the start of a project so that diverse opinions are accommodated and reflected in the number and variety of novel ideas that enter the funnel for the project. However, once a selection has been made of which idea or ideas are going forward, the team should coalesce around that decision and the task conflict, in theory at least, should diminish (Kratzer et al., 2006). Thereafter, the project is entering the next phase which is likely to be (idea) conversion followed by (idea) diffusion or implementation.

Task conflict or role polarity is neither prevalent nor useful at this point. However, in the case of the US, it was precisely at this point that the role polarity began to surface as they struggled to definitively eliminate some of the ideas even after the consumer research had rejected them. This conflict may have contributed to the incremental nature of the portfolio of ideas ultimately presented by the US. Indeed, this is very likely to have happened given the predisposition of one of the team members to oppose any radical ideas which were being evaluated by her team. This member of the US team effectively acted as their resident 'exit champion' as she discouraged the team's adoption of any radical ideas.

In the UK, because they didn't adhere to any form of stage-gate process, role polarity unnecessarily permeated the entire process. In essence, the UK filtered their ideas purely on the passion exhibited by the originator of the ideas and were not able to make any type of market volume estimate.

This case illustrates the importance of developing a scoring mechanism by which to evaluate new ideas within whatever context this is being done. That neither team did this, helps explain why the participants reported such frustration at the process (or lack of it) and at some of the ideas being carried forward. An elementary screening mechanism would have helped the teams identify an optimum balance between the desirability of some of their ideas (in terms of their revenue potential) against the degree of difficulty envisaged in bringing those ideas to fruition; in terms of probability of technical success, how long it might take and the reliance on external partners.

8.4.7 Harnessing the Voice of the Consumer

In this case study, there is no real dispute about the need for consumer input into the innovation process, the conflict arises over the level of such input; its timing and whether or not it is appropriate to have this marketing research work carried out by scientists whose talents and experience might arguably be better deployed elsewhere.

In the literature, the debate on the role of the voice of the consumer in the innovation process continues without definitive resolution with some arguing that consumers are limited in their capacity to develop ideas outside the current use context, and that they are consistently lacking in foresight (Hayes and Abernathy, 1980; Hamel and Prahalad, 1994; Martin, 1995). Proponents of this argument further believe that technology can advance at a much faster pace than some users comprehension of it and hence they will have difficulty in inputting any useful insights in technology driven markets (Moriarty and Kosnik, 1989; Lynn et al., 1996; Veryzer, 1998a; O'Connor and Veryzer, 2001).

However, some researchers assert that there is a group of users – specifically 'lead users' – who can contribute product or category insights based on wants and needs they experience significantly earlier than other more mainstream users and are therefore able to anticipate early trends and possibly even emerging markets or market segments. Thus, some advocates believe that lead users can possibly be a source of ideas which might lead to radical innovation (von Hippel, 1986, 1989; Urban and von Hippel, 1988; Herstatt and von Hippel, 1992; Lilien et al., 2002).

In 'Democratizing Innovation' (2006, p. 19), Von Hippel notes that the idea that new products and services are developed exclusively by manufacturers is deeply ingrained both in individuals' experience and in academic writing. He reflects that when individuals perceive some shortcoming in a product or service, they generally respond by wishing 'they' (the manufacturers) would 'do something about it.' Even the term 'consumers', he suggests implies a passive role of consumption only. Von Hippel (2006) notes that new product development is becoming increasingly user-centred where individual customers are exhibiting a growing tendency to innovate for themselves. In his research, he cites markets, specifically fringe sports like surfboarding or some ICT projects, where the level of NPD attributable to consumer input can reach up to 40%.

The NPD literature has seen a good deal of research about the need to involve customers in the process, from the earliest possible stage. The voice of the consumer or customer needs to be integrated into the entire process in order to fully capture the value of their input (Cooper, 1999; Von Hippel, 1986; Brown and Eisenhardt, 1995; Thomke, 2003). Empirical studies have also found that many of the innovations developed entirely by users do have commercial appeal and prospects. As an example, Urban and Von Hippel (1988) found that an industrial software product concept developed by lead users had greater marketplace appeal than did concepts developed by standard marketing research methods. Holger et al. (2010) asserts that the collection of customer information is critical in the very early stage of NPD (Ottum and Moore 1997) because the level of uncertainty regarding customer requirements is high. Some companies involve customers only at the end of the process, notably in services, and this approach has led to criticism (van Kleef, 2005). Additionally, in the field of banking, Athanassopolou and Johne (2004) demonstrated that project teams who communicated with lead users throughout the process were more successful than those who only communicated with them at the end of the process.

Schum and Lin (2007) suggest that as all innovation projects stem from some analysis of customer needs and wants and hence customer knowledge drives innovation in successful firms. Implicit in their descriptor of market orientation is a commitment to consumer knowledge, understanding and communication from the very beginning of all innovation projects.

In their proposal for a world class NPD process, based on the global PDMA survey data, they (Schum and Lin, 2007) suggest a method for ensuring that the voice of the consumer is ever-present in the process:

All these world best practices innovative companies make use of a cross-functional team in new product development. This provides diverse viewpoints of customer needs, with a minimum of at least two employees from marketing and technical. (p 1614)

However, there is a subtle difference here insofar as the PDMA best practice guidelines are recommending using a proxy for the voice of the customer; they suggest that someone from marketing or sales can represent the customers' viewpoint. But many companies prefer to have a dialogue with the customer themselves rather than simply with their own customer-facing staff. Cooper (2009) has written extensively on the subject of best practice in the area of NPD and notes:

The quest for unique, superior products begins with a thorough understanding of the customer's unmet and often unarticulated needs—through in-the-field, voice-of-customer work. This means that the entire team—technical, marketing, and operations people—interviews and interfaces with real customers/users, and learns their problems, needs, and challenges firsthand. This is quite different from relying on the salesperson or product manager to speak for the marketplace; such information is often filtered, biased, and incorrect. (Cooper, 2009, p. 11)

Advances in technology make it easier to harness the voice of the consumer. Füller (2010) cites the rise of the internet, user-generated content and social networking and the availability of wiki's, terms such as crowd sourcing, co-creation, user innovation, virtual customer integration, and open innovation have become popular. These terms describe a variety of useful roles consumers may play in the previously organisation-led world of new product development and innovation. Consumers are increasingly considered a valuable source of ideas for innovation.

Overall, there are two main criticisms of the practice of engaging consumers in the innovation process. First, it may make lead times longer as customer feedback has to be integrated into the process at each stage (Fang, 2008). Although, this criticism may be mitigated by the higher likelihood of an outcome which is aligned with customer wants and needs. The second reflects the more profound concern that existing customers have a very limited view of the market and its likely future development. Consequently, many ideas emanating from existing consumers are likely to revolve around incremental improvements often centred on price (they want it cheaper) and access (they want it available more easily). In 'The Innovator's Dilemma", Christensen (1997) suggested that technology can move at a faster pace than consumer preference and that, in many cases, consumers don't really know what they want. Hence, allowing consumers be the architects of new product development can actually be counterproductive by reining in more radical ideas and favouring the more sustaining, incremental ones.

In summary, there is some debate in the literature as to the benefits and the role of customer input in the development of (specifically) very new products. While von Hippel (1986, 2006) proposed that users are an important source of new product ideas, others have argued that being too close to customers or being 'customer-led' may prove limiting both for innovation and firm performance (Macdonald, 1995). These and other studies (Christensen, 1997; Ciccantelli and Magidson, 1993; Neale and Corkindale, 1998) form the basis of an ongoing discussion in the literature about whether customers and users inhibit or enable very new or radical product ideas (Connor, 1999; Slater and Narver, 1998, 1999), and if customer input constrains original ideas and leads only to incremental new products. For instance, Callahan and Lasry (2004) have demonstrated how the importance of customer input in new product development changes with product newness.

The literature suggests that consumers are important in all types of innovation but their role is different (and usually less significant) in the development of radical innovation where technology may play a more dominant part than in incremental innovation, which is more likely to be consumer-led.

In the ISF case study, the US team engaged in consumer research to a very high degree. They used focus groups to validate and optimise their own ideas and then took the best performing ideas and put them into volumetric testing. The high level of research equipped them with a winning presentation but it did attract some comments suggesting that their ideas were both incremental and predictable. In contrast, the UK did not manage to arrange any significant piece of research in advance of the final presentation. Instead, they deliberately kept their scientists focusing on the science until the very last minute. As a consequence, although they didn't have all the market research data to back up their ideas, their ideas were described by a number of people on the SLT as more original and radical than the UK ones. This outcome would appear to lend support to the prevailing wisdom in the literature suggesting that higher consumer involvement is more appropriate for incremental innovation than it is for disruptive projects (Christensen, 1997; Ciccantelli and Magidson, 1993).

8.4.8 Using IT in the Early Phases of Innovation

An additional difference in the teams' approach was clear in their use of IT in the management of the project. In the US, the team leader was a strong believer in using software systems to generate, upload, manage and even vote on ideas. The team all subscribed to Think Tank (a software programme) which allowed them to load up ideas regardless of the time of day or geographic location. The software acted as a central repository for the ideas and allowed each team member see other ideas as they were loaded; allowing them to comment on them or add other 'builds' on the ideas if they so wished. They could also cluster ideas under certain broader typologies or platforms and analyse them by various dimensions. The US team used this systems as a central part of their process and yet they did not develop or promote really radical innovations.

The UK did not use any IT systems in managing either the individual ideas or the overall project. Despite, having three team-members located at a separate site, the UK team leader did not subscribe to any specific software to facilitate the generation but, more especially, the sharing of ideas between his team members.

The US team leader made extensive use of specialist innovation software but it was used merely to share ideas internally within his team. It was not used to connect the team or the ideas to external third parties. Ultimately, the team characterised as the incremental innovation team were far more reliant on IT than the radical innovation team. It was the UK's ideas, which were deemed to be more original, novel and radical, and hence one could, within the findings of this case study, question the value of the IT software in terms of its usefulness in developing radical ideas for innovation certainly within team-based initiatives.

8.4.9 Managing Networks in new product development

The case highlights the significance of networks in three important and separate dimensions: internal networks (bridging ties); external networks (open innovation) and innovation champions.

Many studies have shown the alliance networks in which firms are engaged can enhance firm learning, their acquisition of knowledge and innovation (e.g. Ahuja, 2000; Shan, Walker, and Kogut, 1994; Smith-Doerr et al., 1999; Soh, 2003). Networks, including personal networks, are believed to enhance organisational innovation for two main reasons. First, involvement in disconnected networks increases creativity and innovation because they provide the people involved with timely access to diverse information and knowledge (Burt, 1992, 2004). Secondly, networks have a positive impact on trust; they enhance social capital because such structures generate trust, reciprocity norms, and a shared identity, which increase cooperation and knowledge sharing (Coleman, 1988; Portes, 1998). Further, networks imply a level of communication and communication has long been associated with superior performance in innovation (Allen 1971, 1977). Brown and Eisenhardt (1995) suggest that both internal and external networks are connected with success in new product development:

Communication among project team members and with outsiders stimulates the performance of development teams. Thus, the better those members are connected with each other and with key outsiders, the more successful the development process will be. (p. 354)

The literature distinguishes two forms of knowledge network: (1) *contact networks*, through which firms source knowledge; and (2) *alliance networks*, through which firms collaborate to innovate. Networks in the form of the latter (alliance networks) usually describe formalised, purposeful collaboration and joint ventures, or some other 'contracted' relationships resulting in frequent and continuous interaction (Huggins and Johnston, 2010). Specifically, in the area of discontinuous, radical innovation, Birkinshaw et al (2007) have developed a more detailed taxonomy (Figure 8.5) of the types of networks that firms can engage in to help access broader potential sources of innovation projects or simply ideas.

Figure 8.5: Examples of Networks for Discontinuous Innovation

Idea Networks	A set of relationships with individuals and organizations who the firm can tap into to help solve technical problems or to brainstorm new ideas. For example, P&G's Connect and Develop and Eli Lilly's Innocentive.
Corporate Venturing Networks	Involves building relationships with hundreds of prospective new ventures and other VCs with a view to developing a window on new technologies and making selective investments in promising new ventures. For example, Intel Capital Nokia Ventures.
Lead User Groups	A set of relationships with leading-edge customers who help the firm to experiment with and try out new product ideas. For example, Lego's Mindstorm User Group or the BBC's Backstage.com project.
Cross-Industry Alliances	Creation of relationships with various different actors in a particular industry to achieve something that they cannot achieve on their own. For example, Rio Tinto's work with sustainable development agencies on its Breaking New Ground initiative.
Communities of Practice	Cross-boundary and cross-organizational groupings engaged in experience and idea sharing around shared knowledge fields, particularly at the intersection point where two "knowledge worlds" collide. For example, technical groups/knowledge communities at 3M, Xerox, and HP.
Supplier Networks	Networks of partners with whom firms share their strategic roadmaps and invite ideas and inputs to shaping and delivering on new and alternative visions. For example, Rolls Royce and its strategic supplier program.
Open invitation Networks	Networks of self-selecting volunteer partners who organize around a specific project or issue. A recent example was the innovative approach to film financing by Thai-American film producer Tao Ruspoli who invited investors to contribute a dollar (or more) and become associate producers of his next film.

Source: Birkinshaw, Bessant and Delbridge, 2007.

Having networks that extend outside the organisation is referred to as *Boundary Spanning*. Research on boundary spanning generally supports the considerable advantages associated with access to external sources of knowledge and information (Allen and Cohen, 1969; Allen, Tushman, and Lee, 1979; Tushman and Scanlan, 1981; Chesbrough, 2004).

Specifically, it has been noted that boundary spanning plays an important role in driving innovation within; organisations (Cohen and Levinthal, 1990), business units (Hansen, 1999; Tsai, 2001), teams (Ancona and Caldwell, 1992; Reagans and Zuckerman, 2001), and individuals (Burt, 2004; Perry-Smith, 2006). Similarly, other researchers have also shown that such networks reaching outside an organisation are positively correlated to individual (Cross and Cummings, 2003) or unit-level performance (Tsai, 2001). At a conceptual level, therefore, the advantages of boundary spanning are widely supported. Teece (1996) concludes that informal structures of the firm, as well as the network of external linkages that they possess, have a significant bearing on the strength as well as the kind of innovation activity they undertake.

When boundary spanning programmes are extended beyond the idea generation stage to more well developed collaborative projects, closer to exploitation or commercialisation, a number of complications are observed to occur. These arise from the obvious difficulties associated with transferring, integrating, and leveraging the diverse inputs and differing perspectives available across organisational boundaries (Dougherty, 1992; Argote, 1999; Carlile, 2004). However, these difficulties did not become manifest in this project as, with the UK team at least, they did not, in the course of this project, progress the ideas into or close to the activation or implementation phase.

In this case, the use of external networks played an important role in the development of the ideas of the UK team. The team specifically sought external inputs which provided information and connections for the team members. Interestingly, in the case, this use of external networks was on a commercial basis, and was not in any meaningful sense, the result of previous exchanges between the parties. This would suggest that the use of external networks can be a purposeful, targeted and short-term process based on economic exchange. This is somewhat in contrast to the nature of innovation networks portrayed in the literature, where networks are often discussed with reference to the strength of the relationship between the contracting parties.

8.4.10 Innovation Champions

Boundary spanning is most often associated with external communication links, which are critical to enhancing innovations since they facilitate learning and they help secure necessary resources (Goes and Peirk, 1997) and for the successful uptake of ideas between and within organisations (Cziepel, 1975; Daft, 1978; Ghosal and Bartlett, 1987; Kimberly, 1978; Robertson and Wind, 1983). Indeed, Callahan and Salipante (1979) hold that 'boundary spanning units are defined as any group or department whose primary responsibilities are to deal with parties outside the organisation, such as clients, suppliers, and research institutions.'

However, in this case, as the ISF team was a relatively autonomous group within the wider corporate framework, intra-firm boundary spanning was seen to be necessary (by the US team, at least) in order to promote their ideas and to try to secure future support for their development.

One perspective (Youtie and Shapira, 2008) is that in an increasingly knowledge-based environment, high-performing organisations are those which have capacity (through their people in teams), not only to develop, acquire and use new ideas and knowledge, but, just as important, to advance, distribute and recombine this knowledge through the internal promotion of these new ideas. Boundary spanning is a necessary but not sufficient precondition of this activity.

Typically, NPD environments are characterised by high levels of both uncertainty and ambiguity (Mishra and Shah, 2009), therefore, an ability to collaborate effectively both internally and with external partners (Wheelwright and Clark, 1992; Swink, 2006) is central to any organisation's success in the process of new product or service development. Similarly, it is widely recognised that new product development is a highly interdependent process, and successful NPD requires firms to develop routines and practices to collaborate with internal cross-functional employee teams.

Case studies of product innovation success highlight internal championing as a critical means by which social and political pressures are applied to overcome management inertia and help new product ideas find traction within host organisations (e.g., Achilladelis et al., 1971; Schon, 1963; Tushman and Nadler, 1986). Put simply: "A new idea either finds a champion or dies" (Schon, 1963, p. 84).

In the organisational context, raw ideas are potentially very valuable but they are also very fragile and there is an increasing level of academic attention given to theoretical and empirical work on how to ensure the idea survives until appropriate market research can determine whether it should be progressed or abandoned (Brown and Eisenhardt, 1995, Dougherty and Hardy, 1996, Kessler and Chakrabarti, 1996 and Venkataraman et al., 1992). In these studies, the role of a champion generally includes the following inventory of characteristics: active support for innovation, overcoming barriers, attracting resources and ensuring the project makes it to completion (Howell and Higgins, 1990). These have all been shown in research to be vital for the success of any innovation project. (Chakrabati and Hauschildt, 1989; Howell and Higgins, 1990; Roberts and Fusfeld, 1981; Schon, 1963). Glynn (1996) points to the existence of influential "innovation champions" who have the social, political or interpersonal knowledge necessary to accelerate the acceptance of innovative change.
Successful innovation depends, inter alia, on the individual and collective creativity and expertise of employees. Moreover, innovation is increasingly

characterised by an iterative process of people working together building on the creative ideas of one another (Coakes and Smith, 2007). Howell (2005) states that ninety percent of raw ideas never go beyond the originator's desk because they fail to attract a champion (Stevens and Burley, 1997). Dedicated champions, Howell asserts, are pivotal to innovation success and thus ought to find organisational support for their work and be integrated into the mainstream of organisational activity i.e. a champion should not be equated with a maverick.

In recognition of the importance of innovation champions, Howell et al., (2005) developed a three-factor model in which they identified facets of champion behaviour. The key behaviours are: Expressing enthusiasm and confidence about the success of the innovation,
Persisting under adversity, and
Getting the right people involved.

Barczak and Wilemon (2007) see being an innovation champion as an essential trait for an R&D team leader:

Team leaders need to develop, manage, and sustain the team's relationship with senior management. Since disagreements about organisational issues create conflict and affect team members' perspectives about a project, team leaders need to engage in activities aimed at promoting and generating support for the project.

Earlier insights into champion behaviour were developed by Burgelman (1983). Champions articulated a convincing master strategy for the idea, and got support for it informally by "acting as scavengers, reaching for hidden or forgotten resources to demonstrate feasibility" (1983: 238). The next phase for them is establishing and maintaining contact with top management to keep them informed and enthusiastic about the project. Similarly, Venkataraman et al. (1992) observed that a new venture idea required a champion to exert social and political effort to galvanise internal support for the concept and settle conflicts between key stakeholders. Coalition building is another key activity of champions as well as other cooperative influence tactics in order to gain organisational traction for the idea (e.g. Galbraith, 1982, Markham and Griffin, 1998 and Shane, 1994).

While a lot of research on innovation champions attribute much of their positive influence to an inspiring, transformational style of leadership (Chakrabarti and Hauschildt, 1989; Smith et al., 1984; Schön, 1963). Research has also shown that the technical skill and reputation of the champion can be equally important. Technical competence and analytical skills (Beatty and Gordon, 1991) are important assets of an innovation champion. Such technical abilities can be instrumental when it comes to ensuring that the champion has the required and perceived authority within the organisation. Jensen and Jorgensen (2004) suggest that when organisation members regard the champion as one of the leading experts in the organisation in the area of current interest, the champion's chances of breakthrough are considerably enhanced. Analytical skills are also seen as important for innovations where budgeting, planning and control are central tasks and these are areas where the US team leader excelled. Overall, this implies that leadership and managerial skills of the champion are also influential especially when it comes to driving a significant innovation project in the organisation (Beatty and Gordon, 1991; Conger and Kanungo, 1987; Oberg, 1972).

In summary,

'The success of a creative idea is above all a matter of political activities and strategies of the ideator. It pays attention to the notion that organisational creativity is an individual expression on the one hand and organisational commitment on the other' (Bakker et al., 2006. p306).

Additionally, innovation teams not only form the engine and heart of the innovation process (Buijs, 2007), they are also essential for encouraging the organisation to accept and adopt the innovation result. Thus, it is not enough for the innovation champion merely to facilitate their team mates in coming up with a potentially winning idea, he or she also has the difficult task of finding internal support to progress the idea through the innovation process within the wider organisation.

Because the innovation process is a complex, non-linear activity, the role of a champion while widely considered essential for success, is nonetheless difficult to define. Howell et al. (2005) have created taxonomy of champion behaviour that is aligned with prior case study findings and this could make it easier for organisations to facilitate the emergence of hitherto largely informal innovation champions. In the ISF project, it could be argued that a key role of the team leaders was to act as innovation champions, coordinating the efforts of their own teams and facilitating external contacts and networks at the start of the project.

In the ISF project, the US team leader was particularly adept at these champion behaviours. Apart from expressing a lot of confidence in his team and their ideas, he made key senior people aware of the exact nature of the ideas that were circulating around his group and he anticipated any possible objections that might arise. By commissioning the specialist services he did (copywriting, illustration, qualitative focus groups and online volumetric testing), he managed to effectively ward off any possible future criticism that the ideas were too technical and had not been filtered through consumer research. Specifically, he was very adept in both getting the right people involved and keeping the right people updated.

In the case of the UK team, they decided not to attempt to influence senior management by exposing them to the ideas on which they were working. On examination, this seems a naïve approach and one which was not appreciated by the SLT in general. The UK team leader refers to his desire to 'surprise' the SLT with some sparkling ideas. By deliberately not sensitising SLT members to what was coming from his team, Grist inadvertently engineered a situation where his ideas had to work much harder to gain traction with the judging panel.

In this case, success wasn't just judged upon the originality of the ideas or concepts but was considered to depend upon the organisation's subsequent likelihood of adopting the team's ideas. Hence, the US team leader acted as an exemplar of some of the characteristics commonly ascribed, in the literature, to innovation champions and his teams' ideas had already been previewed to the influential judging panel. Ideally, towards the end of the project, the focus of the champion's role might have (and had in the case of Coapman) switched to finding internal future sponsors and building a coalition of support for the ideas. In the case of the US team, this role was discharged to a very high level while in the UK, it was largely ignored. The failure to engage in the coalition building required to seed the ideas with the company pipeline is definitely one of the factors that caused the UK team to lose the competition.

The case data adds some support to the literature's view that innovation champions can have a powerful effect in getting influential support for ideas within the organisation. The evidence from this case is, however, slightly equivocal, because it was the team whose ideas were incremental who exhibited championing behaviours and secured support for the ideas. It could be argued that the company already had an ability to assimilate and a tendency to promote incremental ideas and hence it is hard to attribute this support exclusively to the role of the innovation champion. Nevertheless, the case endorses the view that ideas can travel far further in the organisation if they attract the attention of influential senior managers who can allocate time and resources to them.

8.4.11   The Role of the Team Leader in Innovation Projects

Academic interest in the role of the team leader in R&D projects has been high and sustained for 30 years now. And yet, despite this, a number of authors continue to assert that this is an area, which has hardly been touched in research. Buijs (2007, p. 203) declares there to be a gap in the literature in the area of leadership of innovation teams which he notes 'has hardly been discussed in the innovation literature.' Mumford et al. (2002) note that this topic is 'conspicuously absent' in the literature. Edmondson (2003) merely suggests that the leadership role in R&D has been down-played in the literature while Elkins and Keller (2003) point to 'a lack of theory-based leadership studies' in the R&D environment. Nippa (2006) says:

Comprehensive reviews of the broad research on critical success factors of managing product innovation in most cases do not emphasise leadership or leadership styles explicitly. (p2)

However, since 1980, a number of studies have sought to investigate and characterise this role. The R&D team in any firm, which is responsible for most of the organisation's product innovation, plays, probably, *the* crucial role in firm survival (Huang and Lin 2006). Finding effective methods with which to manage an R&D team so that it achieves a high level of innovation performance should be on top of the agenda of any business (Beheshti, 2004). The literature on this topic is prolific, consistent and reasonably clear about the skills and priorities of R&D team leaders likely to deliver the highest level of performance from their teams.

Many studies agree that the R&D team leader needs to be an exceptional individual, combining both soft and hard skills in equal measure. Just as an innovation project has essentially two elements; invention and implementation (Amabile, 1988; Dougherty and Hardy, 1996; Kanter, 1988; Klein and Sorra, 1996); equally the R&D manager must have the skills to manage those two phases. But each phase requires very different skills not just from the leader but from the team involved as well.

Nippa (2006) points out that the characteristics required in the ultimate R&D leader are not easy to distinguish or describe; 'the search for the perfect R&D leader, i.e. superior characteristics, seems either to be still in the starting-block or up a blind alley.'(p3)

There is a dichotomy that is not fully resolved in the literature. It is this; innovation work relies, in its early stage on an ability to work in the currency of creativity, novelty and ideas (Cardinal, 2001). This requires creativity and risk-taking and a high tolerance for ambiguity. This mainly characterises the 'R' stage of conventional R&D. To lead a project of this type demands that the leader facilitate a creative process that has an element of uncertainty and chaos at its heart. Many studies come to the conclusion that relationship-orientated or transformational leadership styles are superior within the context of innovation generally (Bass, 1985; Jung, 2001; Sosik, Kahai, and Avolio, 1998). This seems intuitively right, that charismatic, open, inspiring leadership would be the approach that would encourage idea generation and development when success, in that phase of the project, depends on the generation of high quality ideas.

But innovation, at least in its latter, implementation phase, also involves process-driven and routine tasks. A key question in this debate is whether one individual can really manage the entire, end-to-end; fluid to fixed; divergent to convergent; flexible to structured; open and experimental to closed, process of innovation. Nippa (2006. p4) sums up the difficulty thus:

The perfect R&D leader (note: on the supervisory level) – whether born or socialised – possesses an optimal mix of superior characteristics and qualities that enables him or her to be everybody's darling and to achieve all objectives for the benefit of all stakeholders. (p 4)

NPD Project Leaders will not always be involved in the creative process themselves but research shows that they exert an enormous influence on the people who are directly involved. To undertake creative work successfully, individuals must have certain abilities; expertise and creative skills such as general cognitive problem solving skills (Reiter-Palmon and Illies, 2004), and task motivation (Amabile, 1997). A supervisor or leader will exert a direct influence on their subordinates' motivation to engage in creative problem solving; they will, equally, exert a considerable influence on the motivation of their team with regard to risk-taking. Additionally, R&D leaders, either deliberately or unintentionally, will shape the work environment arising from various elements of the organisational structure, climate and culture, which they often have authority over (Nippa, 2006). In short, they will control the conditions under which team-level creativity will either flourish or perish.

Amabile and Khaire (2008) suggest that: 'One doesn't manage creativity; one manages for creativity.' (p. 102)

Dolan et al. (2003) also point out that, in innovation, it is especially necessary to develop a style of facilitating' leadership to ensure that the right things happen. They see the essential characteristics as being the capacity to inspire, to articulate a vision and to hold teams of creative individuals together and channel their work. Amabile and Khaire (2008) agree and point out that the leader's job is to map out the stages of innovation and recognise the different skill sets, processes and perhaps even technologies that are necessary to support each phase. Their 'simple' advice to people managing innovation and creativity; 'Know where you are in the game.' (p. 104)

Awareness of the leadership traits and skills required for the team or project management level of R&D leadership has attracted considerable research attention. Consequently, the research canvas is very broad covering such elements as leader traits and characteristics (Barczak and Wilemon, 1989), leader position and power (Ancona and Cadwell, 1992), project climate (Harbone and Johne, 2003), leadership styles (Sarin and McDermott, 2003) and autonomy (Jassawalla and Sashittal, 2000).

Brown and Eisenhardt (1995) confirm that the heart of the product-development process and the focus of much research is the project team.

'Project team members are the people who actually do the work of product development. They are the people who transform vague ideas, concepts, and product specifications into the design of new products.'

Hence the project or innovation team is the very engine of new product or service development. Barczak and Wilemon (2003) suggest:
Senior management needs to ensure that team leaders possess the three sets of critical skills identified by our research: interpersonal, project management and technical. As a first step, management needs to carefully consider whom they place into team leadership positions. (p. 476)

Within the ISF project, in the US, another, separate issue emerged about an individual member of the team who rejected the corporate vision implicit in the ISF mandate. This person was personally sceptical about the organisation's appetite for radical innovation and she acted accordingly. In the academic literature, the characteristics of their organisation largely affect followers' creativity (Siegel and Kaemmerer, 1978; Scott and Bruce, 1994; Amabile et al., 1996). According to Scott and Bruce (1994), organisational climate is an important factor for creativity; employees' perceptions of the extent to which creativity is encouraged at the workplace, and the extent to which organisational resources are allocated to supporting creativity influence creative performance. An employee's perception of an innovative climate encourages risk taking, and the challenge to use creative approaches at work. This employee, clearly, did not perceive the organisational climate for creativity to be genuinely capable of accepting and progressing truly radical ideas and she acted accordingly in trying to block the adoption of the group of ideas outside the current footprint of the organisation.

Possibly the most important decision to be taken in such an innovation project is the selection of team leader. The leader needs to be able to combine components of the archetypal transformational leader when inspiring team members with a compelling vision. He also needs to possess solid technical skills to be able to judge the quality of the team's work and to command the team's respect as an expert in his field. As an expert, the team leader must be able to forge useful networks both within and, especially, outside the organisation. At a minimum, the team leader needs to be positively disposed and open to external collaboration even if she doesn't personally possess the necessary contacts to make it possible. Additionally, she should have a clear view of the direction and pace with which she wants her team to move. This might involve creating sub-teams to combine skills and projects in order to meet the project objectives. The leader also needs to use political acumen and experience in finding a path to take the team's ideas forward within the organisation; a process, which invariably involves enrolling senior level champions to believe in and support the team's ideas.

The leader then needs to possess the transactional skills and style to make things happen and to channel the work towards an agreed output once the creativity has done its job.

In light of this, when reviewing the ISF case, while the approaches and activities of the two team leaders were very different from each other; they each had some elements to recommend them. At a simplistic level, one can imagine that the perfect R&D team leader would be an optimal blend of the finest qualities of the two subjects in this study.

Chapter Nine

Conclusions & Contribution

9.1   Introduction

This chapter has three objectives: first, to outline the contribution this study makes to the field of new product development and innovation and to position this research in terms of its import to policy and practice, second, to discuss the limitations of the research and third, to make recommendations for future research.

9.2   Research Question

The objective of the study was to understand how R&D teams in large organisations generate and develop radical innovation ideas. Despite an increased academic interest in NPD structures, the question of how firms should implement an effective NPD process design for radical innovation remains largely unanswered. Sundstrom and Zika-Viktorrson (2009) acknowledge that although innovation is being studied with ever increasing frequency and intensity, there have been many calls for a better understanding of how innovation really happens in organisations (Goffin and Mitchell, 2010; Grönlund et al., 2010).   Van de Ven and Poole (1990; p.313) hold that 'an appreciation of the temporal sequence of activities in developing and implementing new ideas is fundamental to the management of innovation.'

The research question addressed in this study is as follows: how does innovation happen in a large, complex, global organisation? Specifically, within an R&D setting, how can a firm create a culture, a climate and a team that will deliver radical innovation ideas?   How are such ideas generated and how should they be tested and ranked, and how do organisations decide which ideas to pursue and which to abandon? How are, or should, teams be managed to deliver on a firm's innovation objectives? What organisational factors might encourage innovation, as well as what factors might inhibit innovation?

9.3 Findings

It is widely acknowledged that in terms of new product (and service) development and their impact on both the industry and firm level; significant differences exist between radical and incremental innovations (Dosi, 1982; Christensen and Rosenbloom, 1995; Christensen, 2000). There is a plethora of classifications to define radical innovations (in particular) with some emphasising their impact on the technology; others stressing their disruption to the market; still others focusing on their novelty and more relating the radicalness to its impact on the parent organisation.

Based on an embedded twin-case study within the R&D division of GSK; following the processes and outcomes of two competing innovation or NPD teams a number of factors have been shown to be conducive to the development of radical innovation.

The variation in outputs and outcomes for the teams; with one biased heavily towards radical ideas and the other veering almost exclusively towards incremental ideas was not envisaged a priori in the design of the ISF initiative. It transpired that the team leaders chosen for both teams had vastly diverging views on how to accomplish the objectives and their individual leadership styles was one critical factor that influenced these outcomes. Neither approach was ideal and perhaps the perfect approach might have been somewhere in between the two described.

Chapter Eight provided a response to the research question which asked how does innovation unfold inside large, complex, organisations in the R&D setting; and, what factors might be conducive to the development of radical ideas for innovation and which organisational factors might frustrate radical innovation. Chapter Eight has reprised the key themes both from the perspective of the case itself and then from the standpoint of the literature.

Four factors within the process in the case study emerged as being central to the variation in outputs that characterised the teams: structures, processes, networks and leadership. As part of the analysis, these factors were integrated into a framework (potentially) for application by managers who are charged with the responsibility of either setting up or managing innovation teams and projects within their organisations.

Figure 9.1: Proposed Framework for Managing Innovation Teams & Projects

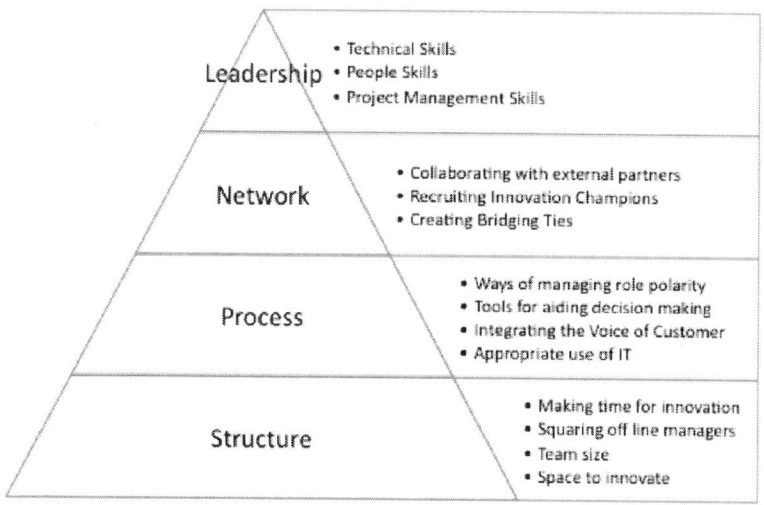

Leadership	• Technical Skills • People Skills • Project Management Skills
Network	• Collaborating with external partners • Recruiting Innovation Champions • Creating Bridging Ties
Process	• Ways of managing role polarity • Tools for aiding decision making • Integrating the Voice of Customer • Appropriate use of IT
Structure	• Making time for innovation • Squaring off line managers • Team size • Space to innovate

9.4 Contributions

As introductory context, it is important to note that this study is set within a natural experiment within a global firm. The case study addresses the overall NPD process but the case data is specifically rooted in the heart of the fuzzy-front-end or the early stage of the innovation process. Hence, the study's contribution will be strongest in guiding managerial practice and scholarly research in the first phase (Idea Generation) of the innovation value chain (Hansen and Birkinshaw, 2007).

This study makes a number of contributions. Whetton (1999) notes that most scholars do not develop an original theory, rather they add something to an already existing theory. In the case of this study, the key area, construct or concept being examined is the practice new product development within firms.

A theory is defined as a collection of assertions or propositions that identify what variables are important and for what reasons. It specifies how they are inter-related and why, and it identifies the specific conditions under which they should or should not be related (Campbell, 1990). Straw and Sutton (1995) asserted that theories generally answer questions of 'why?' This perspective was echoed by Christensen and Sundahl (2001; p 2) who concluded that; 'A theory specifies what causes what, and why, and under what circumstances.'

In this study, as in a number of others (Leifer et al, 2001), a pragmatic approach has been taken to defining radical and incremental innovation; i.e. the former being innovation which, for the promoting company, requires both new technology and new marketing capabilities because it means entering new markets or new market segments. Thus, radical innovation represents an opportunity for firms which offer both higher risk and potentially higher reward.

As Phillips et al (2006) confirmed, radical innovation is not a new concept, major upheavals in technology or markets have been happening since before the Industrial Revolution. The issue they emphasise for managers is that the events or trends that trigger such upheavals; like the advance or adoption of new technology, the fragmenting of markets, the rising power of the consumer and regional instability in political and economic systems are all on the increase. Hence, managers will be faced with a situation where radical innovation needs to be proactively planned for and facilitated to an unprecedented degree.

The contributions of this research are as follows. First, it is argued that the method and context of the study is, in itself, a contribution to the literature on innovation. Specifically:

- In terms of method, the study sought to connect the inputs to the innovation process to the outcomes that were developed within the project teams. This is in contrast to much of the existing literature on innovation management which tends to treat these two aspects in a somehow separate way (Isari and Pontiggia, 2010). That is, researchers often focuses either focus on how the process of innovation is organised in firms (innovation models) or, alternatively, on the results of the innovation processes and their characteristics (innovation outcomes).

- The study is based on a revelatory case study. As noted in in Chapter Two, the majority of contributions to the leading journal on innovation are quantitative in nature, with case based research in the minority; with deductive theorising being the most prominent approach to research in organisations in general (Shepherd and Sutcliffe, 2011) and to innovation studies in particular. Thus, in its methodological approach, this study acts as a counterpoint to the prevailing orthodoxy of positivist research.

- But, most importantly in this section, an aspect of context worth noting is that the design of the innovation initiative, which constitutes the case study, is unique in this organisation and certainly rare in any organisation. The ISF programme itself marked a cornerstone of the reorganisation of GSK's R&D division. This ISF programme underlined the organisation's commitment to innovation; its objective was to liven up the NPD pipeline by generating radical ideas. This project design lends itself to types of observation, comparison and analysis, which would be impossible with more conventional cases. This rare access into the real workings of an innovation team and project operating at a very high level in one of the world's biggest pharma companies, builds on previous pharma R&D case studies (Balsano et al., 2008; Bonabeau et al., 2008). Additionally, while

ISF was an organisational experiment separate from and outside other innovation programmes being concurrently run by the company; it also had a second experiment nested within it. This second experiment refers to the choice of the two team leaders; one a process expert and the other a scientist.

Second, this study suggests a set of factors and a style of leadership that might be associated with radical innovation.

Third, the study suggests that there is a differential potency associated with each variable connected with radical innovation, with the impact depending on the stage of the project along the innovation value chain (IVC). Many prior studies have treated the innovation process as a homogenous unit and have neglected to sub-divide it into its very separate component parts. Such a division is made in the literature around the stage-gate process but this is mainly to align mechanistic, to-do lists with each stage of the process. This research, on the other hand, is suggesting that the factors previously connected with radical innovation are of increasing or decreasing relevance to managers depending on the phase of the IVC in which each specific innovation project is situated.

For example, the contribution of individual creativity to an overall radical innovation project depends on the stage at which the project is at, with arguably its strongest contribution coming at the idea-generation stage. Similarly, the contribution of strong project management capability is likely to have a beneficial effect throughout the project but this effect is likely to be greatest at the latter, implementation, diffusion stage of the project.

By considering how stage of the innovation life cycle might influence the nature of the innovation process, the study contributes to the debate on the question of if, how and when the voice of the consumer (VOC) or customer should be included in the innovation process. Extant literature suggests that it is unclear if the VOC is an advantage to an innovation team or whether it limits the variety and scope of ideas that are considered. This study suggests that the inconclusive nature of this debate may reflect the failure by some of the extant research to differentiate between the stages of the innovation process and the nature of the innovation sought. So, the case data suggests that during the earlier ideation phase when radical ideas are the desired outcome, it may be advantageous to delay the VOC input to the process.

Fourth, extant literature also presents the innovation process in terms of a stage-gate process. This study suggests that when radical innovation is the quest for teams, the usefulness of stage-gate processes may be limited. These mechanistic approaches to innovation can have the effect, if they are too rigidly applied, of stifling creativity and killing promising ideas too early. Moreover, stage-gates represent the phases of the process to be equivalent in length whereas the case data suggests that spending longer in the idea generation phase and involving more (internal and external) people in that part of the process is more likely to yield radical ideas for innovation.

Fifth, this study addresses an important gap in the literature regarding the importance of leadership styles to the performance and outputs of innovation teams (Ancona and Caldwell, 1992a; Barczak and Wilemon, 1992; McDonough and Barczak, 1991; McDonough and Griffin, 1997; Sarin and McDermott, 2003; Sarin and O'Connor, 2009). Prior literature has explored leadership in R&D and innovation but studies typically fail to distinguish between the type of leadership that is appropriate or required according to the phase of the innovation process. The case suggests that style of leadership is contingent on the stage of the innovation process. Specifically, the case suggests that loose and flexible leadership may be associated with radical innovation. The converse also holds, with tight management being associated with incremental innovations.

Fourth, this study contributes to our understanding of the role of networks in innovation. Existing literature suggests that collaborating with organisational and individual networks contribute to a likelihood of success with radical innovation (Karkkainen and Ojanpera, 2006; Caswill and Wensley, 2007; Bahemia and Squire, 2010), while the closed innovation system is correlated with incremental innovation. More recently, Steiner (2009) suggests that collaborative creativity (i.e. tapping into external sources of ideas) is a prerequisite for the generation of ideas for innovation that aims at radical instead of incremental improvements of products, processes and services. Steiner (2009), in his call for 'open creativity', notes the particular contribution a networked, collaborative approach can make in the creative, idea-generation phase of innovation: 'In a way, what "open creativity" is for creativity, Chesbrough's "open innovation" is for innovation.' (Steiner, 2009, p. 5). The association between external networks and radical innovation is evident in this case data. The UK team, whose activities included external connections with experts, innovation intermediaries, universities and research institutes, developed a portfolio of radical ideas. The US team whose idea-generation activities were principally confined to desk research produced mainly incremental ideas. The data extends existing literature on the role of networks by suggesting that the recruitment of external partners to collaborate in innovation projects may in some cases be relatively easy and swift to effect. This suggests that it some regards it may be relatively straightforward for some organisations to switch from a closed to an open innovation model of innovation.

9.5   Limitations of the Research

The qualitative nature of this research, with its focus on a single organisation and twin case study, means that the findings cannot be confirmed by traditional, statistical measures of reliability. The nature of qualitative research generates outcomes that cannot be viewed as facts or objective truths; they are the result of an interaction between the research and the researched (Silverman, 2000). Thus although these frameworks cannot be represented as complete theories in themselves, they do highlight issues of interest to those involved in the academic and practitioner community.

A further limitation is that not only is this case study embedded within one organisation; it is also limited to one industry sector, the healthcare or pharmaceutical industry.   This industry is characterised by a significant investment in and focus on R&D to an extent that is unlikely to be matched in other industries that are not so research intensive.

An element of bias may be inherent in this case study because of the fact that it represents insider research. The selection of an embedded case study design (in GSK) poses a particular challenge arising out of the researcher's close involvement with the organisation.   Coghlan and Brannick (2005, p. 61) argue that insider-research facilitates 'the knowledge, insight and understanding of organisational dynamics, but also the lived experience of one's own organisation.' All of which are difficult to replicate with much legitimacy as an outsider with merely cursory familiarity of the context.

Nevertheless, in the current research, it was a particularly important consideration and challenge to maintain objectivity and neutrality.   This was partly accomplished by relying on multiple informants; by including interviewees from different levels within the organisation as well as integrating extensive evidential support from documents and archives in order to support the study's findings.

Conventional definitions of innovation give equal weight to both invention and implementation (Bessant and Tidd, 2011). Hansen and Birkenshaw's (2007) model extends this innovation definition into a three-stage journey; idea generation; idea conversion and idea diffusion. This project revolved mainly around the invention or idea generation element with the implementation phase mainly devoted to building a coalition of future support to find champions to promote the ideas internally. Hence, some of the elements found to be associated with the successful delivery of radical innovation are of limited relevance here as they apply more to the implementation or diffusion phase; a phase which played a more minor role in this instance.

These limitations suggest that the results have to be interpreted with a slight degree of caution. However, the chosen methodology in the current study contributed significantly to the trustworthiness of the results. Therefore, the results are interesting and significant for both researchers and managers. While the results are in no way generalisable, they are potentially transferable to organisations of similar scale and context to the one described. Thus, the value of the findings, in terms of their generality, results from their degree of credibility to those with an interest in the research area (Wyatt, 2001).

9.6   Implications for Practice and Policy

This research is of value to managers who participate in, use, commission or manage teams to deliver on NPD or innovation objectives as it indicates potential project or team-level factors that may facilitate or hamper the likelihood of innovation success. It identifies a range of contingencies, which potentially have a bearing on the appropriate calibration of the structure, processes, networks and leadership for innovation teams. In summary, it provides the following broad guidelines or recommendation to managers involved with or leading innovation projects:

Structure

1) Choose the team carefully and ensure that participants have sufficient time to engage with the innovation programme by adjusting their workload and objectives according to their expected input to the project.

2) Be mindful that the team will perform best when there are not too many members; 7 or 8 may be an ideal number.

3) Ensure that the appropriate line managers have been briefed and that the necessary flexibility in reporting arrangements has been anticipated.

4) Space to innovate: assign an appropriate creative space in which the team can meet and where they can store and interact with project material

Processes

5) Ideally, have access to a commercial expert in order to assist in the prioritisation of the ideas by using market data

6) Develop an outline framework for prioritising the early-stage ideas which is flexible enough to allow for radical ideas but sensitive enough to be able to distinguish between potentially major and minor ideas

7) Ensure that the entire team are aligned behind the objective and deal with any outlying opinions which may impede the team's work

8) Plan for the appropriate integration of the voice of the customer in the project, while acknowledging that how this is done may differ with each type of idea

9) Be judicious in the use of innovation IT and software, they may make managing the project easier, but they are unlikely to be the catalyst for radical ideas.

Networks

10)   Ensure the team is connected to all the relevant stakeholders and experts within the organisation so that they can leverage any tacit knowledge that may be available.

11)   Connect the team with external networks of research institutes; universities; innovation intermediaries and consultants so that third party ideas are included in the ideation phase.

12)   Begin to position the ideas with senior management and identify champions for them to facilitate their subsequent adoption into the pipeline after the innovation team has handed them over.

Team Leader

13)   Choose a team leader who has the necessary skills and experience to facilitate the process through the innovation value chain.

A further point; the ISF and this subsequent analysis of the project and its outcomes offer a rich learning opportunity for GSK. Specifically, it allows them the opportunity when considering similar initiatives to integrate the lessons learned from this project. The author has been invited to present these findings to the organisation in 2011. The framework proposed is merely a beginning; it needs testing and refinement. It is hoped that GSK might adopt this model in future innovation programmes within the R&D group and this will allow the model to be tested.

As noted in the Introduction, Ireland faces an innovation imperative; new product and service development is the central pillar of current (2010) government economic policy. Apart from public service innovation over which it has some control, the government can only achieve its ambitions for innovation through enhancing the innovation capacity of indigenous Irish businesses. These insights and managerial recommendations will assist Irish companies to develop and manage their own innovation projects, possibly with a greater likelihood of success. The Community Innovation Survey (CIS, 2008) indicates that only 47% of firms in Ireland are 'innovation active' which means that the majority of Irish firms are not engaged in innovation or new product and service development. One of the most often cited reasons for this absence of focus on innovation within Irish firms is 'lack of knowledge or specialist training in innovation'.

Managerial guidelines and insight around the management of innovation, such as the issues raised and the frameworks proposed here are likely to be of considerable interest to Irish businesses as they look for practical methods to engage with new product and service development.

9.7   Suggestions for Future Research

While the research examined a single company in a single industry and this limits the generaliseability of these results, this study has provided in-depth insights which can be extended to future research.   For practitioners, it underlines the importance of linking structures, processes, networks and leadership in innovation projects and of maintaining feedback mechanisms that help adapt these elements over time.

This study addresses a number of questions in the research on new product development.   It is, however, rooted upstream in the early phases of the process and its import is most pronounced at the first phase of the innovation value chain. Hence, it suggests some questions for a research agenda which extend further along the value chain towards implementation.   Among these questions is what happens to the outputs of these experiments inside the organisation as they get handed off to other managers and teams and begin the next phase of their life in the NPD pipeline. How are they assimilated into the organisation as they move closer to commercialisation. Are there any factors likely to provide them with a higher likelihood of success as they enter the commercial funnel.

Within this experiment, one notable feature was that there was little evidence of the guiding hand of corporate strategy. The project teams were merely asked for a quota of ideas across a number of different therapy areas as well as a few ideas outside the company's current business footprint. Future research might examine how strategy, including portfolio strategy might help to give such programmes strategic clarity and organisational connection.

Further research would be helpful in establishing a correlation between the resources, including time, devoted to the early stage ideation and the eventual novelty of the innovation outcomes. It seems logical that there should be a positive connection between the amount of time devoted to the idea-generation phase of innovation and the quality of the ideas emerging from it. However, this is possibly a curvilinear relationship which reaches a point after which any additional time spend ideation may actually be counterproductive.

Another area of future research surrounds organisational creativity. This study illustrated a very high level of creativity from the UK team which appeared to flourish under conditions of extremely loose management. It is noted in the literature that innovation is generally thought to be a group activity while invention is an individual act (Amabile 1998). In the UK team, there was a high level of invention and relatively little group activity. A useful area of future research might be to investigate the conditions most conducive to individual creativity in the context of innovation teams in particular. How individual creativity gets channelled into organisation and team level initiatives merits further attention.

A key question which was not explored in this study is how firms make the transition back to operations once they have developed a radical innovation. Sometimes, existing business units within the firm will either be uninterested or inappropriate as the operating home for the new offering. In these cases new SBU's (strategic business units) may be created or, alternatively, the initiative can be spun-out to a third party for commercialisation. Each of these possibilities has repercussions for the ultimate success of the venture and it would make an interesting research study to compare the various routes to operational commercialisation of truly radical innovations that emerge from large organisations.

Another potential future research stream concerns the handling of the personnel involved in the development of a successful radical innovation project. Inevitably, they will pass the baton to a business manager. The receiving unit generally wants to assimilate the new product or service into existing operations. Leifer et al, (2002) refer to this handover as the 'transition gap' and note that it is beset with difficulties. They find that not only does successful transition require the active input of people from both sides of the gap, but it is often also necessary to have the process facilitated by people who are expert in managing these transitions. Such tensions are likely to be most acute in radical innovations and some template for managing the transition would be beneficial to managers.

This case study was investigating a firm in the healthcare business and this also restricts the generaliseability of the findings. Future research could therefore examine the nature of innovation practices in industrial or technology firms to extend our understanding of those industries. This would go some way in addressing whether the typology advanced here is relevant in a wider context.

Future research efforts could also be applied to empirically testing this innovation framework

9.8   Final Comment

The subject of innovation is one of increasing importance at government level, at sector level, at industry level and at company and project level. While decades of research have been able to show that innovation is no longer a dark art, nor is it solely attributable to luck – it is still seen as something of a managerial black box whose secrets have yet to be fully revealed.

Uncertainty plagues radical innovation projects. The first, and often greatest, uncertainty is the technical one; can we make this? Will it do what we claim for it? These are quickly followed by market uncertainties such as who will want to buy it and are there enough buyers to make it a worthwhile venture? But equally frustrating, in the background are often resource and organisational uncertainties such as will the project be funded this year? Or how can we find more sponsors and champions to believe in this project? Aligned to these uncertainties are issues around managing senior management expectations for radical projects and devising appropriate metrics for ideas that are genuinely novel.

Paradoxically, all this uncertainty and risk makes radical projects the most exciting innovation initiatives to be involved in. After all, if they are successful, the payoff is dramatic for the firm. This excitement was what attracted me to embark on my study. The case study contained in this thesis, I believe, offers an interesting insight into how radical innovation can be encouraged inside large organisations. Given the long gestation period involved in the NPD process in healthcare, often requiring clinical trials to demonstrate efficacy and safety, a number of the projects initiated within this project are still actively being developed by GSK.

5769965R00243

Printed in Great Britain
by Amazon.co.uk, Ltd.,
Marston Gate.